Ecological Modelling

Ecological Modelling

An introduction

S.E. Jørgensen
University of Copenhagen, Denmark

WIT*PRESS* Southampton, Boston

S.E. Jørgensen
University of Copenhagen, Denmark

Cover images by Mette Vejlgaard Jørgensen

Published by

WIT Press
Ashurst Lodge, Ashurst, Southampton, SO40 7AA, UK
Tel: 44 (0) 238 029 3223; Fax: 44 (0) 238 029 2853
E-Mail: witpress@witpress.com
http://www.witpress.com

For USA, Canada and Mexico

WIT Press
25 Bridge Street, Billerica, MA 01821, USA
Tel: 978 667 5841; Fax: 978 667 7582
E-Mail: infousa@witpress.com
http://www.witpress.com

British Library Cataloguing-in-Publication Data

A Catalogue record for this book is available
from the British Library

ISBN: 978-1-84564-408-6

Library of Congress Catalog Card Number: 2009932146

The texts of the papers in this volume were set individually by the authors or under their supervision.

Printed in Great Britain by Athenaeum Press Lld.

Contents

Preface ix

Chapter 1
Ecological models **1**

1.1 Why go modelling? 1
1.2 History of ecological modelling 3
1.3 Modelling components 6
1.4 Modelling procedure 8
1.5 Selection of model complexity 19
1.6 Model uncertainty 21
1.7 Mediated modelling 22

Chapter 2
Model types **29**

2.1 Classification of models 29
2.2 Model types – an overview 29
2.3 Characteristics of the model types 33
2.3.1 Dynamic biogeochemical and bioenergetic models 33
2.3.2 Steady state biogeochemical models 34
2.3.3 Population dynamic models 34
2.3.4 Structurally dynamic models 35
2.3.5 Fuzzy models 35
2.3.6 Artificial neural networks 36
2.3.7 Spatial models 37
2.3.8 Individual based models and cellular automata 37
2.3.9 Ecotoxicological models 38
2.3.10 Stochastic models 39
2.3.11 Hybrid models 39
2.4 Applicability of the different model types. 40

Chapter 3
Biogeochemical models **45**

3.1 Biogeochemical and bioenergetic models have wide applications 45
3.2 A chemo-state model to illustrate a steady state biogeochemical model. 48
3.3 An eutrophication model based on 2–4 state variables and statistic correlations to find additional state variables 49

3.4 An eutrophication model of medium to high complexity 54
3.5 Model of subsurface wetland 66
3.5.1 Process equations 69
3.5.2 Parameters 70
3.5.3 Differential equations 71
3.5.4 Model results 72
3.5.5 Practical information about forcing functions and parameters 73
3.5.6 Use of the model for wetland design 73
3.6 River models 73

Chapter 4
Population dynamic models 87

4.1 Introduction 87
4.2 Basic concepts 87
4.3 Growth models in population dynamics 88
4.4 Interaction between populations 93
4.5 Matrix models 99
4.6 Metapopulation models 101
4.7 Modelling infectious diseases 102

Chapter 5
Ecotoxicological models 109

5.1 Introduction 109
5.2 Five classes of ecotoxicological models 111
5.2.1 Food chain or food web dynamic models 111
5.2.2 Static models of the mass flows of toxic substances 111
5.2.3 A dynamic model of a toxic substance in one trophic level 112
5.2.4 Ecotoxicological models in population dynamics 113
5.2.5 Ecotoxicological models with effect components 113
5.3 The application of parameter estimation methods in ecotoxicological modelling 116
5.4 Biogeochemical and ecotoxicological model example: contamination of an agricultural field 120
5.5 Uptake of toxic substances by plants 125

Chapter 6
Spatial models 133

6.1 Introduction 133
6.2 2D and 3D models 134
6.3 An example of distance as independent variable: chromium pollution in a Danish Fjord 135
6.3.1 Model description 136
6.4 Application of GIS 141
6.5 Use of cellular automata or lattice models in spatial modelling 141
6.6 Surface modelling 143
6.7 Use of IBM in research: determination of the role of conjugation in the evolution 144
6.7.1 The model 145

6.7.2 State variables 146
6.7.3 Model structure 147
6.7.4 Parameters and forcing functions 147
6.7.5 Model results 148
6.7.6 Discussion and conclusions 150

Chapter 7
Structurally dynamic models (SDMs) 153

7.1 Basic concepts of structurally dynamic modelling 153
7.2 Ecosystem characteristics 154
7.3 The thermodynamic concept of eco-exergy 159
7.4 Structurally dynamic models 163
7.5 Development of SDM for Darwin's finches 168
7.6 Model of the ectoparasites–birds interactions 169
7.7 A structurally dynamic eutrophication model developed by use of STELLA 170
7.8 SDM developed for Lake Fure 173

Appendix 1: Table of oxygen solubility as function of temperature and salinity 185

Index 187

Preface

Ecological Modelling: an Introduction is a textbook tailored for the last year of a bachelor study or the first year of a master study. Ecological ModeLing is a very powerful tool in environmental management and ecological research and a course in ecological modelling ought, therefore, to be compulsory for all studies focusing on environmental management, environmental sciences and ecology. Ecological modelling requires both an introduction to the basic theory and practical computer exercises in the development of models. To facilitate the exercises, the book presents several illustrations where an ecological modelling problem or task is presented and the solution is then given in detail. Several exercises are presented after each chapter, but the best exercises are of course that the students develop their own models focusing on problems that they have formulated and have particular interest to solve. The basic theory of ecological modelling is covered by the first seven chapters of the textbook, including 'how to go modelling' and an overview of the various model types applicable in ecological modelling. The following five chapters present details about the development and application of the five most applied model types in ecological modelling: bio-geo-chemical models, population dynamic models, ecotoxicological models, spatial models and structurally dynamic models. For all five types, detailed and illustrative examples are presented. The examples are chosen to give the readers sufficient information to facilitate the development of their own models of the same type.

Sven Erik Jørgensen
Copenhagen 2009

CHAPTER 1

Ecological models

1.1 Why go modelling?

Living organisms are able to make models. The more complex the organism is the more complex models it is able to make. Mankind has always made more or less descriptive models of the environment, which have often been used to solve problems or to obtain knowledge about how nature is functioning. When we use mathematics as a useful tool in our description of nature, we talk about mathematical models in contrast to physical models, which try to imitate the physical appearance of the modelled object, for instance, a model of a ship.

Models give a simplified picture of reality and will never contain all the features of the real system because then it would be the real system itself. It is, however, important that the model contains the characteristics of the system that are essential in the context of the management or scientific problem to be solved or comprehended.

Geographical maps, which are models too, illustrate clearly the properties of models. We have different types of maps, which are used for different purposes, for instance, maps for aeroplanes, cars, railways, ships, archaeological sites, site-specific geological characteristics and so on. The situation is similar for ecological and environmental models. For instance, there are lake models that are used to manage the eutrophication or acidification of lakes, lake models that give information about a sustainable fishery or lake models that focus on the oxygen profile and so on. Geographical maps have different scales depending on their use and the available data. If we want to find our way in a town, we would like to have maps that show all the streets, while it is sufficient to have a map of the motorways, if we just want to go by car from Copenhagen to Paris. Similarly, for ecological models, if we want to manage the average eutrophication of a large lake, it is probably not necessary to distinguish between the level of eutrophication in different parts of the lake, which, however, would be necessary if we want to ensure a maximum eutrophication level in all parts of the lake.

Let us summarize the comparison of geographical maps with models – particularly ecological models – based on the following points:

1. different scales for different purposes;
2. different types for different purposes;
3. a map or a model cannot contain all detailed information;
4. they are useful simplifications of reality, which is useful in a given context.

Newton's laws of gravity may be considered models. They are also based on simplification, for instance, a constant velocity presumes in most cases no friction, the falling apple has no air resistance and the mass is constant and so on. In spite of the simplifications, Newton's laws are very useful in numerous mechanical calculations, because the simplifications introduce only minor errors that we can determine by the use of other physical laws about friction and air resistance. The required accuracy of our calculations determines whether the simplifications are allowed or whether we should make more complicated calculations considering the friction and the air resistance.

With regard to the use of ecological models in ecology and environmental management, we are in a similar situation. We are dealing with a much more complicated problem than Newton's falling apple and therefore in most cases we have to make several simplifications. The required accuracy determines whether the simplifications are acceptable or not. If they are not acceptable, we have to revise the model with additional processes and components, which may require more data – not different from our use of Newton's laws for mechanical calculations.

As mentioned above, ecological systems are usually characterized by a much higher complexity than physical systems. Fortunately, today we have very powerful computers that can help us develop complex models. However, a crucial question in the development of ecological models is: "Which are the important processes and components in the model context?" To answer this question, we need to know the problem and the ecosystem. Simplifications in mechanical calculations and in the development of ecological models require that we can distinguish the essential processes and components from the minor important ones, which should not be included in the model, because they only make the model unwieldy and difficult to overview.

Our problems are getting more and more complex and more and more comprehensive due to, among other factors, globalization. Therefore, it is becoming more and more difficult to survey all the factors that may influence the problems. Science is using an analysis of the problem and separates often in the laboratory a problem and a few factors having influence on the problem; but many of the real-life problems are too complex to be solved by this scientific method. A synthesis of many factors and their interactions is needed and it implies that we have to develop a model to be able to survey the many possibilities. Some of the interactions and processes may be known from science; but for real-life problems including environmental problems there are many interactions operating simultaneously. Fortunately, models have been developed to deal with the high complexity. Today, we have wide experience in the development of ecological models. Many thousands of models have been published during the last three decades. The journal *Ecological Modelling* today publishes more than 4000 pages per year. Ecological and environmental models are able to support strongly our decisions in ecological and environmental management.

So, building a model helps us to perceive the interconnections and to connect previous experience with present observations and simulate what would happen in the future. Model development often requires a multidisciplinary approach, because processes and interconnections from many disciplines may be an integrated part of the problem. The last several decades' experience with the development of ecological models has shown us the clear advantages that models are able to offer ecology and environmental management today [1]:

- Models are the synthesis of all what we know: observations, theoretical knowledge, knowledge about rates and sizes, knowledge about food items, etc.
- Models are tools to provide an overview of complex systems.
- Models make it possible to quantify by the use of mathematical formulations and computers.

Therefore, models give a much more profound picture of the problem and the system than, for instance, a statistical treatment of observations, because a model can include *all* our theoretical knowledge.

1.2 History of ecological modelling

The history of ecological modelling forms six stages. The emergence of the first still applied ecological models, the Lotka–Volterra Model and the Streeter–Phelps Model, in the 1920s form the first stage. These models were developed further during the second stage. The Streeter–Phelps Model was applied for other aquatic ecosystems and integrated with hydrology in this stage. Several mathematical stability analyses and modifications of the Lotka–Volterra Model were performed during the 1950s and 1960s. These analyses were used to make population dynamic predictions that were at least partially supported by observations.

The real take-off in ecological modelling was, however, the third stage. This phase of modelling is characterized by the:

1. use of computers as powerful tools in modeling;
2. development of more complex models, which became possible due to the application of computers;
3. use of ecological models to solve environmental management problems, as illustrated in Fig. 1.1.

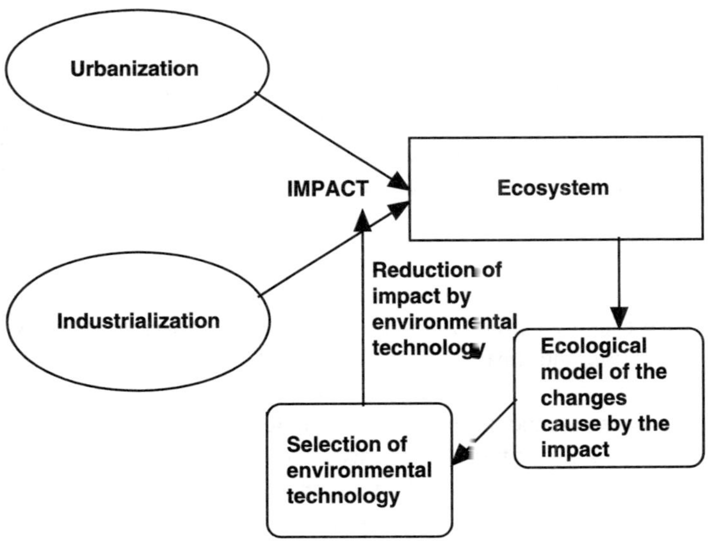

Figure 1.1: The figure illustrates the idea behind the application of ecological models for environmental management, which started in the early 1970s. Industrialization and urbanization have impact on the ecosystems mainly due to discharge of pollutants. The ecological models are able to relate the impact and the subsequent changes in the ecosystems. It is now possible to go from acceptable changes to acceptable impacts. The selection of environmental technology was made based on the impact reduction needed to obtain the acceptable changes in the ecosystem.

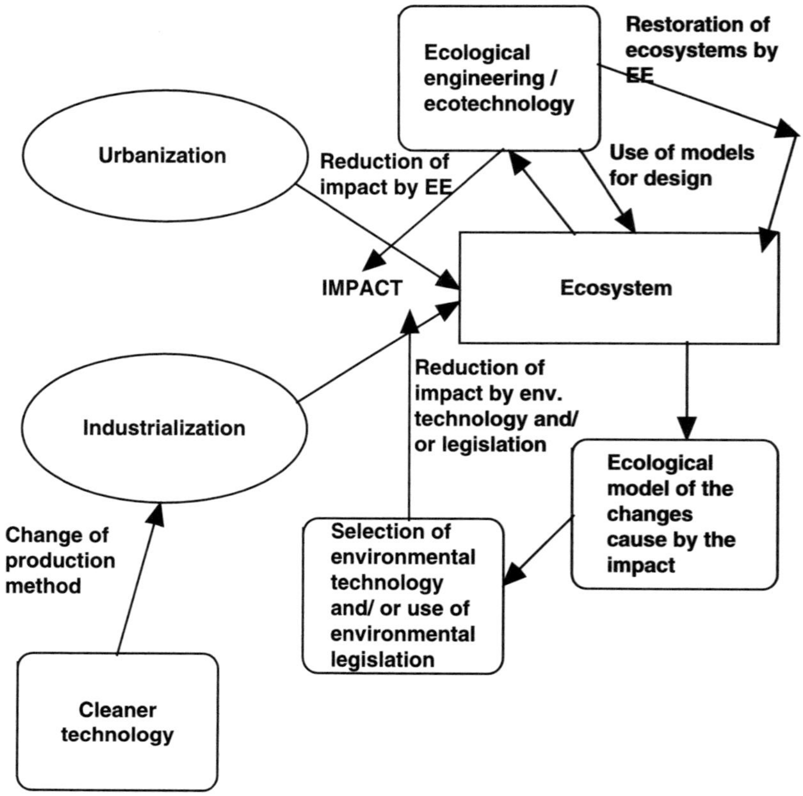

Figure 1.2: Today, models are applied more widely than 35 years ago, because ecological models are used to select environmental technology, environmental legislation, ecotechnology and cleaner technology. The application of ecotechnology also requires the use of models for the design and restoration of ecosystems.

Today, models have found an even wider use and it is also possible to apply ecotechnology, environmental legislation and cleaner technology in addition to environmental technology to achieve the needed reduction of the impact on ecosystems (see Fig. 1.2). The acceptable impact is found from the ecological model, because an ecological model gives the relationship between the impact and the subsequent changes in the ecosystem. When we decide which changes we can tolerate, the model can find the corresponding acceptable impact.

The first eutrophication model was published by Chen in 1970 [2]. The development of several ecological models focusing on different types of ecosystems and different environmental problems gave birth to a new ecological subdiscipline, ecological modelling.

A scientific journal covering this field was launched in 1975. In the first year, it was hardly possible to fill one volume or 320 pages with model papers. In 1975, 40 papers were published and in 1980 the number of papers published was 81. The journal has since grown steadily. Today the journal publishes about 4000 pages/year and the number of papers submitted to *Ecological Modelling* is about 2.5 times as much – or expressed differently, the rejection rate has to be about 60% to correspond to the 4000 pages.

Ecological modelling matured during the fourth stage, covering the period from 1975 to 1980. It became clear from the experience gained by the development of many new models that it was necessary to follow strictly a stepwise modelling procedure with several tests of the model towards the available observations. It also became clear that ecosystems are very different from physical systems and that the characteristics of ecosystems should be reflected in the model [3, 4]. During the fourth stage, the field of ecological modelling grew in depth.

During the fifth stage, a number of different ecological models – of toxic substances in the environment and of almost all types of ecosystems and environmental problems – were developed.

During the fifth stage, ecological modelling grew in scope. The ultimate test of a model, that is, validation of a previously published prognosis, was performed on a few models with partially good results during the early 1980s.

The sixth stage covers the period from about 1990 to today. During this period, a number of new model types found wider application, for instance, individual based models (IBMs), artificial neural networks (ANNs), structurally dynamic models (SDMs), fuzzy models and spatial distribution models. These models are all presented briefly in the next chapter. The journal *Ecological Modelling* expanded in this period from less than 100 papers per year to about 380 papers per year during the last few years. The sixth phase is also characterized by an increasing number of models for terrestrial systems, which are usually more difficult to model than aquatic ecosystems, due to their heterogeneity.

Originally, papers on ecological economics, ecological engineering and ecological indicators were accepted and published in *Ecological Modelling*. Today, these topics are published in separate journals. As a ramification of the journal *Ecological Modelling – Systems Ecology*, the following journals have been launched: *Ecological Economics* in 1988, *Ecological Engineering* in 1992, *Ecological Indicators* in 2001, *Ecological Complexity* in 2004 and *Ecological Informatics* in 2006. Papers published in these journals could also have been published previously in *Ecological Modelling*; but today ecological economics, ecological engineering, ecological indicators, ecological complexity and ecological informatics are new subdisciplines in ecology.

These new subdisciplines including ecological modelling and systems ecology, totalling seven subdisciplines, are all characterized by using quantitative methods and by being applied to a certain extent in environmental management or to understand the underlying theory of ecosystems. The increasing importance of ecological modelling as a subdiscipline in ecology and the increasing use of ecological models as a powerful research and environmental management tool are also reflected in the citation index of the journal *Ecological Modelling*. The citation index has increased steadily and for the year 2008 it was 2.18.

With the increasing power of computers, it has been possible to develop more and more complex models. The models published in the 1970s did only rarely consider the spatial distribution of the model components. Today, several 3D models of aquatic ecosystems have been developed. The development of the geographical information system (GIS) has also made it possible to represent in a very illustrative way the spatial distribution of the considered model variables. GIS is increasingly applied for the development of ecological models, as it can be seen from the increasing number of GIS-based ecological model papers that have been published during the last decade. Several landscape models integrating several different ecosystems have also been developed during the last decade.

In environmental management, there is a need for the development of social–economic–ecological models, because the social–economic implications of environmental management decisions have to be taken into account. Some first attempts to develop social–economic–ecological models have been taken, but it will probably take 8–12 years before we have gained sufficient experience in the development and use of these very complex models. In about 10 years, it is expected that these models will be widely applied in environmental management.

Ecological models must build on good knowledge to the reactions and processes of ecosystems. It became clear in the 1970s that it was important to consider the ecosystem properties in the development of ecological models. Systems ecology is therefore important for ecological modelling, and ecological modelling can be used as a tool in system ecological research. Fortunately, systems ecology has developed in parallel to ecological modelling. Today, we have been able to integrate several contributions to an ecosystem theory that can be used to explain ecological observations and rules (see [5–7] and *Ecological Modelling*, 156(4), December 2002). The presented ecosystem theory is of course the first attempt and it will, as all scientific theories, be modified and improved during the coming years.

To summarize:

1. Ecological modelling is used increasingly in both research and environmental management.
2. Many models of almost all types of ecosystems and for almost all environmental problems are available today.
3. The complexity of the developed models has increased. Landscape models have been developed successfully and many models presenting the spatial distributions have been presented.
4. Further development of ecological modelling will integrate socio-economic problems. It is foreseen that this type of integrated model can be used in management in about 10 years.
5. We do have an ecosystem theory that can be used as an additional tool in the development of ecological models. It is foreseen that there will be a continuous exchange of ideas between ecological modelling and systems ecology also in the coming years.

1.3 Modelling components

Models have five types of components.

1. Forcing functions, or external variables, are functions or variables of an external nature that influence the state of the system. The problem to be solved can often be formulated as follows: If certain forcing functions vary, how will this variation influence the state of the ecosystem? The model is used to predict what will change in the system, when forcing functions are varied with time. The forcing functions under our control are often called control functions. The control functions in ecological models can, for instance, be inputs of toxic substances to the ecosystems and in eutrophication models the control functions are inputs of nutrients. Natural forcing functions in contrast to control functions are, for instance, the climatic variables, which influence the biotic and abiotic components and the process rates.

   We distinguish between the ecosystem and its environment, that is, the entire world minus the ecosystem. Forcing functions account for the influence of the environment on the ecosystem.
2. State variables describe, as the name indicates, the state of the system. The selection of state variables is crucial to the model structure, but the choice is often obvious. If, for instance, we want to model the bioaccumulation of a toxic substance in an ecosystem, the state variables should be the organisms in the most important food chains and the concentrations of the toxic substance in the organisms. In eutrophication models, the state variables will be the concentrations of nutrients and phytoplankton. When the model is used in a management context, the values of state variables predicted by changing the forcing functions can be considered the

results of the model, because the model contains relations between the forcing functions and the state variables.

3. Mathematical equations are used to represent the biological, chemical and physical processes. The total set of equations describes the relationship between the forcing functions and state variables. The same type of processes may be found in many different environmental contexts, which implies, of course, that the same equations can be used in different models. This does not imply, however, that the same process is always formulated by use of the same equation. First, the considered process may be better described by another equation because of the influence of other factors. Second, the number of details needed or desired to be included in the model may be different from case to case due to a difference in complexity of the system and the problem. Some modellers refer to the description and mathematical formulation of processes as submodels. A comprehensive overview of submodels can be found in the ecological literature. See, for instance, Chapter 3 in *Fundamentals of Ecological Modelling* by Jørgensen and Bendoricchio [1].
4. Parameters are coefficients in the mathematical representation of processes. They may be considered constant for a specific ecosystem or part of an ecosystem. In causal models, the parameter will have a scientific definition, for instance, the excretion rate of a toxic substance from a fish. Many parameters are not indicated in the literature as constants but as ranges, but even that is of great value in the parameter estimation, as will be discussed further in the following text. A comprehensive collection of parameters in environmental sciences and ecology can be found in the literature [8, 9]. Our limited knowledge of parameters is one of the weakest points in ecological modelling, a point that will be touched on often throughout this book. Furthermore, the applications of parameters as constants in our models are unrealistic due to the many feedbacks in real ecosystems. The flexibility of ecosystems is inconsistent with the application of constant parameters in the models. A new generation of models that attempts to use parameters varying according to some ecological principles seems a possible solution to the problem, but further development in this direction is absolutely necessary before we can achieve an improved modelling procedure reflecting the processes in real ecosystems. This topic will be discussed further in Chapters 2 and 7.
5. Universal constants, such as the gas constant and atomic weights, are also used in most models.

Mathematical models can be defined as formal expressions of the essential elements of a problem in mathematical terms. The first recognition of the problem is often verbal. This may be recognized as an essential preliminary step in the modelling procedure, which will be treated in more detail in the next section. The processes that link the state variables and the forcing functions are, however, easily expressed as mentioned above by the use of mathematical equations. In the most applied type of ecological models – dynamic biogeochemical models – differential equations are applied to express the changes in the state variables. These equations express simply, that accumulation = inputs – outputs, where inputs and outputs are processes that are expressed by algebraic equations.

Many ecological researchers have discovered how powerful ecological models are to test scientific hypothesis about ecosystem reactions. Ecosystems are complex systems and therefore, in most cases, it is not easy to make direct experiments with entire ecosystems. It is, however, possible to make changes, for instance, in the forcing functions and observe the reaction of the ecosystems as changes in the state variables. In such cases, it will be necessary to develop a model of the focal ecosystem with the involved forcing functions and state variables to be able to assess whether the observations are in accordance with a proposed theory or hypothesis. These possibilities have opened up for a wider use of ecological models in ecology, particularly systems ecology, of course.

1.4 Modelling procedure

Figure 1.3 presents a procedure that is often applied in model development. This procedure can be considered the experience gained by the development of several different ecological models during the 1970s. The open question at that time was: How can we ensure that the model is represented as correctly as possible from the integrated knowledge about the system and the problem?

As seen, the first step in all model making is to define the problem and even that would often require a multidisciplinary team, because it is important to quantify all the sources of the problem. The system is defined together with the problem because the problem affects either a man-made or a natural system. For ecological models, it is important to identify the type of ecosystem: lake, reservoir, estuary, wetland and so on. It makes it possible to apply the general knowledge about what is normal for this type of ecosystem. It should be possible at this stage to list the natural and man-controlled forces that are driving the system. As we would include only the most important forcing functions, it may be useful to grade the forcing functions, for instance, on a scale from 1 to 5: 1 – insignificant; 2 – of minor importance for the system; 3 – has some impact on the system; 4 – has a significant impact on the system; 5 – is crucial for the system and its reactions. Based on the grading, it will be possible to determine the forcing functions that should

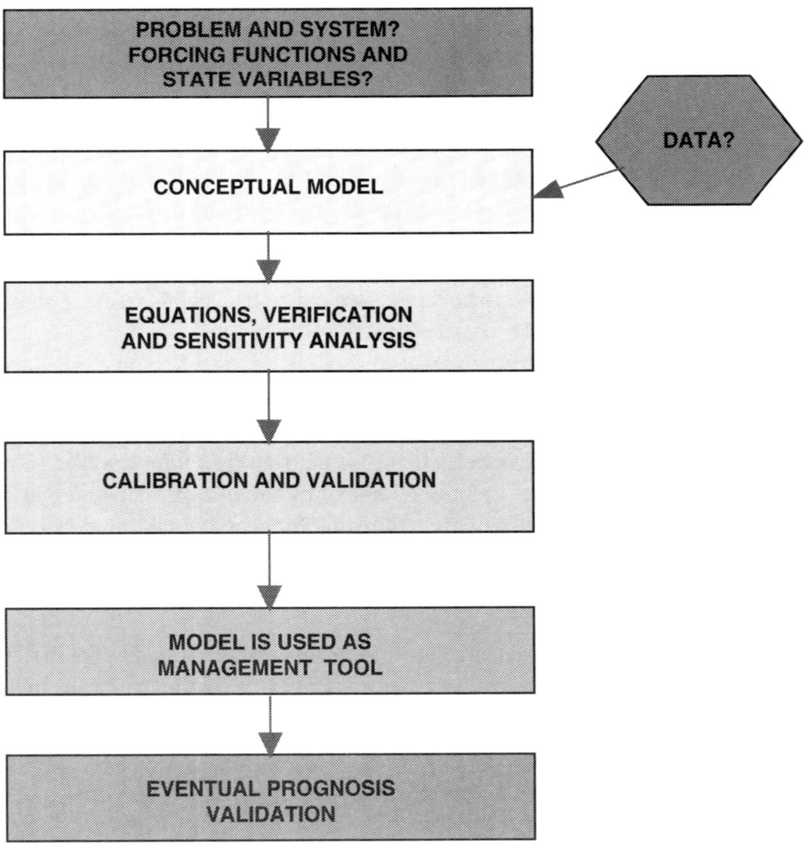

Figure 1.3: A tentative modelling procedure.

be considered in the model. Usually the grade 3–5 forcing functions should be included, while grade 2 forcing functions could be included at least in the initial phase.

It is beneficial to use a mass or energy balance to choose the most important components, forcing functions and state variables, to be included in the model. Let us illustrate the considerations with an example. Let us anticipate that it is an open question, whether birds should be included in a eutrophication model as a state variable. Birds contribute to the inputs of nutrients by their droppings. If the nutrients – nitrogen and phosphorus – coming from the birds' droppings are insignificant compared with the amount of nutrients coming from drainage water, precipitation and wastewater, inclusion of birds both as a state variable and as a contributing component (forcing function) in the model is an unnecessary complication that would only contribute to the uncertainty. There are, however, a few cases where birds may contribute as much as 25% or at least more than 5% of the total inputs of nutrients. In such cases, it is of course important to include birds as a model component or at least as an important forcing function. A mass balance is needed to uncover the main sources of the problem – in the example, the inputs of nutrients.

If a component is participating in an important biogeochemical cycle, it could be important to include the component as a state variable under all circumstances to give a complete description of a cycle of importance for the reactions of the system to the important forcing functions. If observations were available for the component, it would, in most cases, be recommendable to include the component as a state variable. After these considerations, it will, in most cases, be relatively easy to make a list of the state variables. It could also be beneficial for the state variables to use a grading similar to the one proposed above for forcing functions. Generally, it is better at this stage to include slightly too many state variables than too few and it is, in most cases, easier to exclude a state variable at a later stage due to its insignificance for the problem and the system.

The problem, of course, cannot be solved unless the processes, reactions and the interconnections in a real system are considered. In this context, it will be beneficial to ask such questions as:

1. Which biogeochemical cycles occur in the system?
2. Which elements are involved?
3. How is the problem interfering with the cycles?
4. Which of the possible linkages between state variables are actually realized? To answer this question the modeller should go through all linkages. If the model has x state variables the number of possible linkages is $x^2 - x$ as all state variables in principle could be linked to all other state variables. Illustration 1.1 shows how an adjacency matrix can be applied to set up this list.

This does not imply that all processes associated with the problem and the system should be considered during model development. At an early stage of model development, it is important that it is made clear to all members of the model team that the focus is the system – not the details, remembering that the system is more than sum of its parts. On the other hand, it is important to include in the model all the important sources to the problem for the ecosystem. The sources may be internal or external. In the first case, a good description of the processes involved is needed. In the second case, a good knowledge of the forcing functions as a function of time is needed. The selection of model complexity will be discussed in more detail later in this chapter.

In this first step, the problem and the system – for ecological models it is an ecosystem, several interacting ecosystems or parts of an ecosystem – are described by the use of words. The verbal model is, however, difficult to visualize and it is, therefore, conveniently translated in the second step of the modelling procedure into a conceptual diagram, which contains the state variables, the forcing functions and how these components are interrelated by processes that at a later stage in the model development should be formulated by mathematical equations.

As shown in Fig. 1.3, the conceptual diagram is constructed by considering the available data. Due to the calibration of the model (step 4), it is generally recommended to include only state variables that have observations. In a few cases, it may be possible to include one or two state variables that are not observed, if it is difficult to get the observations and it is obvious how the non-observed state variables are related to the other observed state variables.

Figure 1.4 (reproduced from [1]) illustrates a conceptual diagram of the nitrogen cycle in a lake. The state variables are nitrate, ammonium (which is toxic to fish in the un-ionized form of ammonia), and nitrogen in phytoplankton, zooplankton, fish, sediment and detritus.

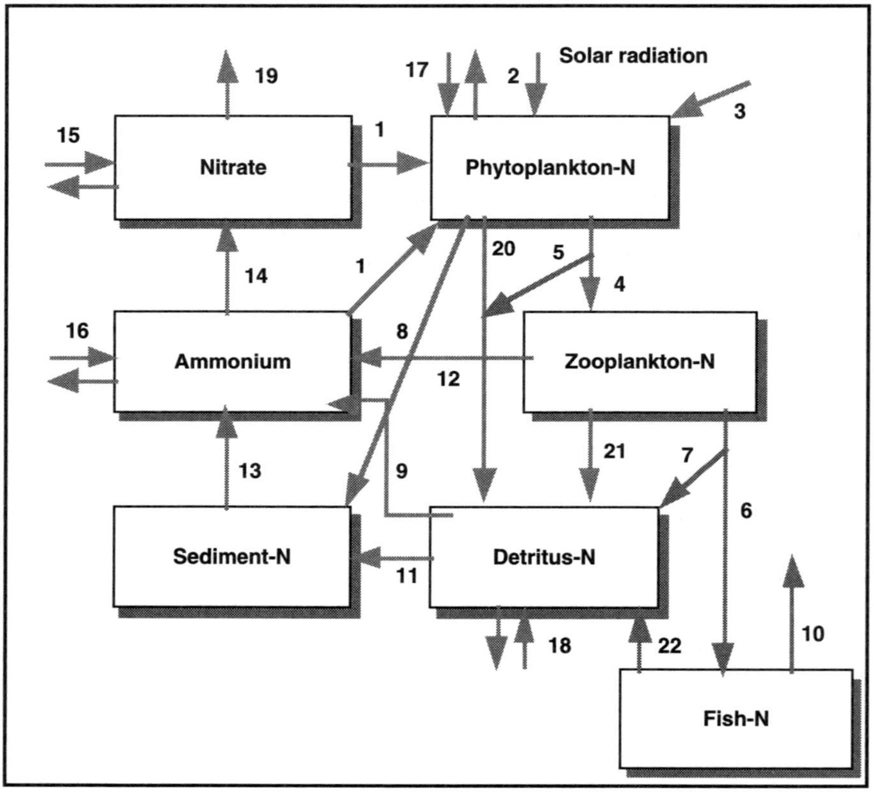

Figure 1.4: The conceptual diagram of a nitrogen cycle in an aquatic ecosystem. The processes that connect the state variables and forcing functions are: 1 – uptake of nitrate and ammonium by algae; 2 – photosynthesis; 3 – nitrogen fixation; 4 – grazing with loss of undigested matter; 5, 6 and 7 – predation and loss of undigested matter; 8 – settling of algae; 9 – mineralization; 10 – fishery; 11 – settling of detritus; 12 – excretion of ammonium from zooplankton; 13 – release of nitrogen from the sediment; 14 – nitrification; 15, 16, 17 and 18 – inputs/outputs; 19 – denitrification; 20, 21 and 22 – mortality of phytoplankton, zooplankton and fish.

The next step comprises of the mathematical formulation. A differential equation is formulated for each state variable: accumulation = inputs – outputs. The accumulation for the state variable S is dS/dt, while the inputs and outputs are processes, expressed by algebraic equations. In the conceptual diagram of the nitrogen cycle in Fig. 1.4, for instance, all the 22 processes shown should and could be expressed by an equation.

The conceptual diagram in Fig. 1.4 indicates the state variables as boxes (e.g. nitrate) and the processes as arrows between boxes (e.g. process number 1). The forcing functions are symbolized by arrows to or from a state variable (e.g. 15 and 16). It is possible to use other symbols, of course, for the modelling components.

The software STELLA, which will be used several times to illustrate the development of models throughout this book, uses boxes for the state variables, thick arrows with a symbol of a valve for the processes, thick arrows coming or going to a cloud for the forcing functions (require a constant, an equation, a table or a graph), a thin arrow to indicate the transfer of information and a variable (forcing function, parameter and/or a state variable calculated by an algebraic expression from another state variable and so on) (see Fig. 1.5).

There are other symbolic languages for the development of conceptual diagrams, for instance, Odum's energy circuit language. It has many more symbols than STELLA and is, therefore, more informative but of course also more time consuming to develop. For an overview of the

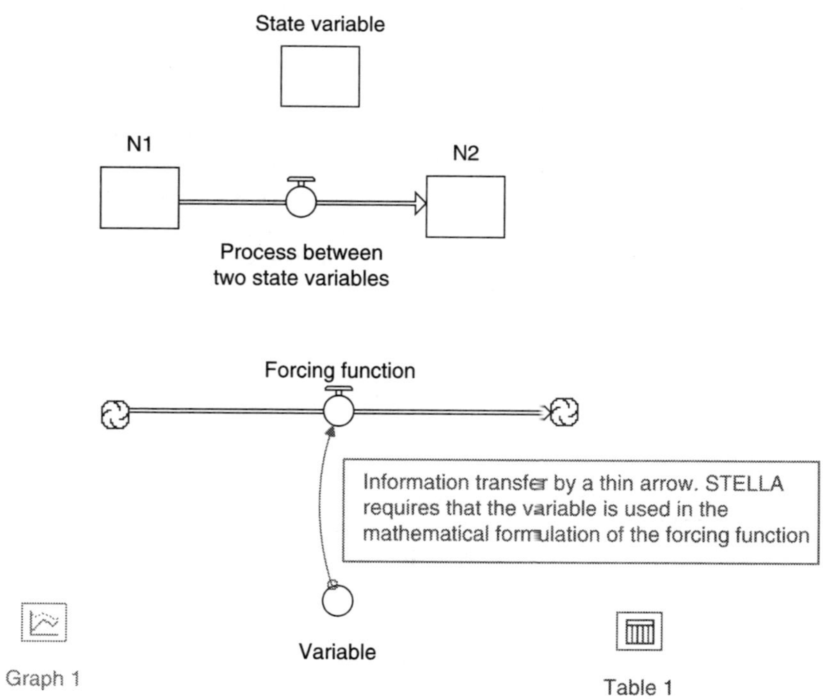

Figure 1.5: The symbols applied to create a conceptual diagram using STELLA. State variables are boxes for which differential equations are formulated as accumulation = inputs – outputs. Processes are thick arrows with the symbol of a valve. Forcing functions are thick arrows starting or ending as a cloud. Circles are variables in general. Graph 1 and Table 1 indicate that the results can be presented as graphs or as tables.

most applied symbolic languages including Odum's energy circuit language, see the book by Jørgensen and Bendoricchio [1]. In this book, we will use either a simple conceptual diagram as in Fig. 1.4 or the STELLA diagram as in Fig. 1.5.

Figure 1.6 illustrates the idea behind the use of differential equations for the state variables. In mathematics, differential equations are solved analytically, while computers solve the equations numerically. A time step is selected for the model calculations by computers. The shorter the time step the closer will the computer calculations be to the real-time variations of inputs and outputs. It is recommended to test different time steps and use the longest time step that would not give any significant change of the model results by decreasing the time step. Significant changes are, of course, evaluated relative to the accuracy of the observations that are used as basis for the development of the model.

The software STELLA formulates the differential equations from the diagram. The time derivative of the state variables will be equal to all the inputs = all process arrows going into the state variables minus all outputs = all process arrows going out from the state variables. The processes must, however, be formulated as an algebraic equation.

Table 1.1 gives an overview of the most applied process equations.

1. A constant flow rate, also denoted zero order expression.
2. A first order rate expression, where the rate is proportional to a variable, for instance, a concentration of a state variable: rate = $dC/dt = k \times C$. This expression corresponds to exponential growth. The following expression is obtained by integration

$$C(t) = C_0 \times e^{kt}.$$

   First order decay has the rate = $dC/dt = -k \times t$ and $C(t) = C_0 \times e^{-kt}$.
3. A second order rate expression, where the rate is proportional to two variables simultaneously.
4. A Michaelis–Menten expression or Monod kinetics known from kinetics of enzymatic processes. At small concentrations of the substrate, the process rate is proportional to the substrate concentration, while the process rate is maximum and constant at high substrate concentrations, where the enzymes are fully utilized. The same expression is used when the growth rate of plants is determined by a limiting nutrient, according to Liebig's minimum law.

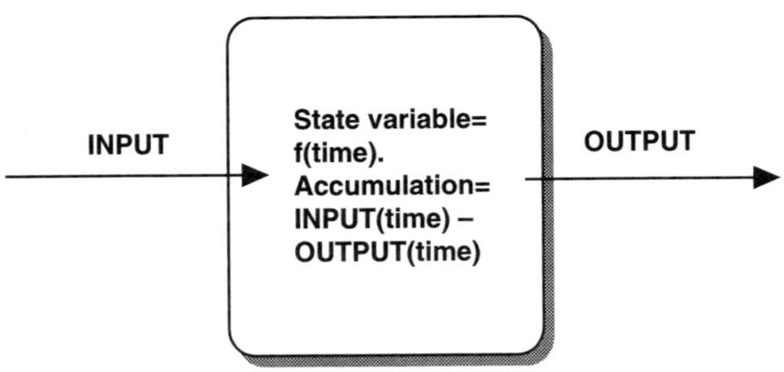

Figure 1.6: The idea behind the use of differential equations is illustrated. The differential equation accounts for the increase of a state variable due to what is denoted in the figure accumulation for a selected time step. The accumulation is input – output. Theoretically, in mathematics the time step is infinitely small, but when the equation is solved numerically by the use of computers, we are selecting a time step. The shorter the time step, the closer is the solution to the mathematical solution.

Table 1.1: Process rate equations.

Expression	Mathematical formulation	Application of the process rate expression
Zero order	Rate = k = constant	Supply is constant f. inst. in film kinetics
First order	Rate = k × variable	Decomposition of organic matter, decay of radioactive components, exponential population growth
Second order	Rate = k × conc1 × conc2	Chemical and biochemical reactions, where the process is based on a reaction of two different components
Monod kinetics	Rate = k × st.var./(st.var. + k_m)	Plant growth by limiting nutrient (Liebig's law) and growth of organisms or populations limited by the food source, f. inst. grazing
Logistic growth	Rate = k × st.var. × (1 – st.var./ carrying capacity)	Growth, where the rate at high values of the st. var. is regulated by another regulation factor
Conc. gradients	Rate = k × dC/dx	Diffusion processes

A graph of this expression is shown in Fig. 1.7. The so-called Michaelis–Menten constant or the half-saturation constant corresponds to the concentration that gives half the maximum rate. At small concentrations of substrate or nutrients, the rate is very close to a first order rate expression, while it is close to a zero order rate expression at high concentrations. Notice also that the rate is regulated from a first order to a zero order expression more and more as the concentration increases.

5. A first order rate expression with a regulation due to limitation by another factor, for instance, the space or the nesting areas. It is expressed by the introduction of a carrying capacity. The general first order expression is applied and regulated by the following factor: (1 – concentration/ carrying capacity). When the concentration reaches the carrying capacity, the factor becomes zero and the growth stops. This process rate expression is denoted logistic growth and it is illustrated in more detail in Section 4.1, including the graphic differences between exponential growth by a first order reaction expression and logistic growth. These two growth expressions are often applied in population dynamic models.
6. Rates governed by diffusion often use a concentration gradient dC/dx to determine the rate, as expressed in Fick's laws: rate = k × (dC/dx) (Fick's First Law).

There are several modifications of these expressions. For instance, a threshold concentration trc is often used in the Michaelis–Menten expression. The concentration (state variable) is replaced by concentration – trc. The concentration, therefore, has to exceed trc to generate any rate. For grazing and predation, it is often used to multiply the Michaelis–Menten expression by (1 – concentration/carrying capacity), similar to what is used in the logistic growth expression. It implies that when the food is abundant (concentration is high) another factor determines the growth, for instance, the space or the nesting area. These modifications will be used in Chapter 3 for the development of a eutrophication model.

Table 1.1 gives information about the applications of these six expressions: for which processes these equations are most often used. Most processes of ecological relevance are covered by these six expressions, although there are, of course, a few processes that will require another

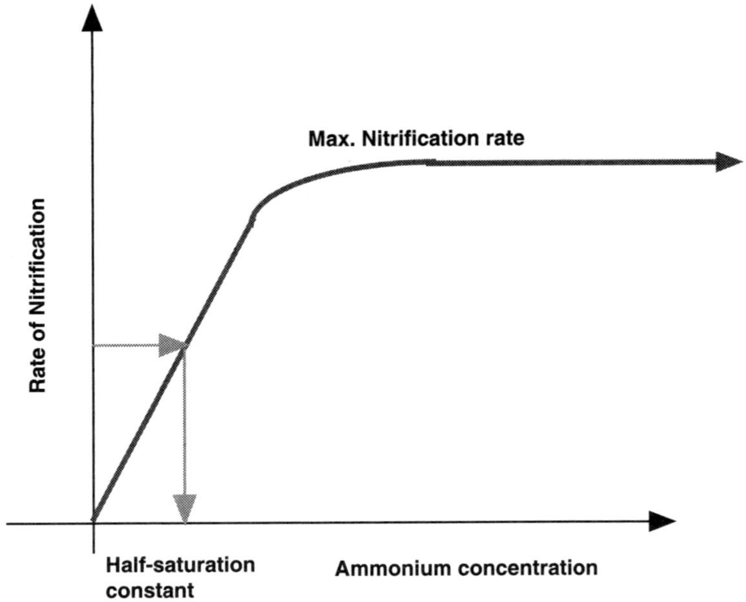

Figure 1.7: The graph illustrates the Michaelis–Menten expression used here for the rate of nitrification; it is dependent on the ammonium concentration. The rate is plotted versus the ammonium concentration.

mathematical formulation. More than 90% of ecologically relevant processes in modelling may presumably be covered by these equations.

The third step of the modelling procedure includes verification and sensitivity analysis. Verification is a test of the internal logic of the model. Typical questions in the verification phase are: Does the model react as expected? Is the model stable in the long term? Does the model follow the law of mass conservation? Is the use of units consistent? The last question implies that all equations should be checked for a consistent use of units.

Verification is to some extent a subjective assessment of the behavior of the model. To a large extent, the verification will go on during the use of the model before the calibration phase.

Notice that during verification it is possible to perform "Gedanken Experiments" similar to how Einstein tested his scientific work. We can, for instance, test a eutrophication model by its reactions to the following experiment: We rent a chopper and buy 10,000 kg of phosphorus fertilizer and drop it instantly into a lake. By use of a model, the experiment can be carried out at no costs, whereas it would be very expensive to rent a chopper and buy 10,000 kg of fertilizer. This is another advantage of models: it is easy to perform "Gedanken Experiments".

Sensitivity analysis follows verification. Through this analysis, the modeller gets a good overview of the most sensitive components of the model. Thus, sensitivity analysis attempts to provide a measure of the sensitivity of parameters, forcing functions or submodels to the state variables of greatest interest in the model. If a modeller wants to develop a model that is able to predict the toxic substance concentration in carnivorous insects as a result of the use of insecticides, he will obviously choose this state variable as the most important one, may be besides the concentration of the toxic substance concentration in plants and herbivorous insects.

In practical modelling, the sensitivity analysis is carried out by changing the parameters, the forcing functions or the submodels. The corresponding response on the selected state variable is observed. Thus, the sensitivity (S) of a parameter (P) is defined as follows:

$$S = [\partial x/x]/[\partial P/P], \tag{1.1}$$

where x is the state variable under consideration.

The relative change in the parameter value is chosen based on our knowledge of the certainty of the parameters. If, for instance, the modeller estimates the uncertainty to be about 50%, he will probably choose a change in the parameters at ±10% and ±50% and record the corresponding change in the state variable(s). It is often necessary to find the sensitivity at two or more levels of parameter changes, as the relationship between a parameter and a state variable is rarely linear.

A sensitivity analysis makes it possible to distinguish between the parameters, forcing functions and processes that have a significant impact on the system behavior and those with a minimal impact on the system. Obviously, the modeller must concentrate his effort on improvements of the parameters and the submodels associated with the high-sensitive parameters, forcing functions and processes.

A sensitivity analysis of forcing functions is in principle not different from the sensitivity analysis of parameters, except that forcing functions are often indicated as table functions of the time or by an equation. The table values or the equation is multiplied by factors that consider the possible uncertainty. If the uncertainty is estimated to be from, for instance, 10–50%, the factors 0.5, 0.9, 1.1 and 1.5 are used and it is recorded how much these factors would change the focal state variable. Based on these results, it is possible to determine how accurately we have to know the forcing functions as a function of the time for a required certainty of the focal state variable.

A sensitivity analysis of submodels (different process equations) can also be carried out. Then the change of a state variable is recorded when the equation of a submodel is deleted from the model or changed to an alternative expression, for instance, with more details built into the submodel. Such results may be used to make structural changes in the model. For instance, if the sensitivity shows that it is crucial for the model results to use a more detailed submodel, this result should be used to change the model correspondingly. The selection of the complexity and the structure of the model should therefore work hand in hand with the sensitivity analysis.

The fourth modelling step encompasses calibration and validation. The scope of calibration is to improve the parameter estimation. Some parameters in causal ecological models can be found in the literature, not necessarily as constants but as approximate values or intervals. To cover all possible parameters for all possible ecological models including ecotoxicological models, we need, however, to know more than 10 billion parameters. It is therefore obvious that in modelling there is a particular need for parameter estimation methods. This will be discussed further in Chapters 3 and 5. Under all circumstances, it is a great advantage to give even approximate values of the parameters before the calibration gets started, because it is much easier to search for a value, for instance, between 1 and 10 than to search between 0 and +∞.

Even where all parameters are known within intervals either from the literature or from estimation methods, it is, in most cases, necessary to calibrate the model. Several sets of parameters are tested by calibration and the various model results of state variables are compared with the observed values of the same state variables. The parameter set that gives the best agreement between model results and measured values is chosen.

The need for the calibration can be explained by the use of the following characteristics of ecological models and their parameters:

1. Most parameters in environmental science and ecology are not known as exact values. Therefore, all literature values for parameters [8, 9] have a certain uncertainty. Parameter estimation methods must be used when no literature value can be found. It is important particularly for ecotoxicological models. In addition, we must accept that parameters are not constant, as mentioned above. This point will be discussed further in Chapters 2 and 7.
2. All models in ecology and environmental sciences are simplifications of nature. The most important components and processes may be included, but the model structure does not account for all possible details. To a certain extent, the influence of some unimportant components and processes can be taken into account by the calibration. This will give values for the parameters slightly different from the real but still unknown values in nature. The difference may partly account for the influence from the omitted details.
3. By far, most models in environmental sciences and ecology are "lumped models", which implies that one parameter may represent the average values of several species. As each species has its own characteristic parameter value, the variation in the species composition with time will inevitably give a corresponding variation in the average parameter used in the model. Adaptation and shifts in species composition will require other approaches as touched on. This will be discussed in more detail in Chapter 7.

A calibration cannot be carried out randomly if more than a couple of parameters have been selected for calibration. If, for instance, 10 parameters have to be calibrated and the uncertainties justify the testing of 10 values for each parameter, the model has to be run 10^{10} times, which of course is an impossible task. Therefore, the modeller must learn the behavior of the model by varying one or two parameters at a time and observing the response of the most crucial state variables. In some (few) cases it is possible to separate the model into several submodels, which can be calibrated approximately independently. Although the calibration described is based, to some extent, on a systematic approach, it is still a trial-and-error procedure.

However, procedures for automatic calibration are available. This does not mean that the trial-and-error calibration described above is redundant. If the automatic calibration should give satisfactory results within a certain frame of time, it is necessary to calibrate only 4–8 parameters simultaneously. In all circumstances, it will become easier to find the optimum parameter set, the smaller the uncertainties of the parameters are, before the calibration gets started.

In the trial-and-error calibration, the modeller has to set up, somewhat intuitively, calibration criteria. For instance, you may want to simulate accurately the minimum oxygen concentration for a stream model and/or the time at which the minimum occurs. When you are satisfied with these model results, you may then want to simulate the shape of the oxygen concentration versus time curve properly, and so on. You calibrate the model step by step to achieve these objectives step by step.

If an automatic calibration procedure is applied, it is necessary to formulate objective criteria for the calibration. A possible criterion could be to find the parameter set that yields the smallest standard deviation between modelled and observed values of all the state variables or a number of selected state variables. Often, the modeller is most interested in a good agreement between model output and observations for one or two state variables, which are considered the main model results, while he is less interested in a good agreement with the other state variables. The modeller may therefore choose weights for the various state variables to account for the emphasis he puts on each state variable in the model. For a model of the fate and effect of an insecticide, the modeller may put emphasis on the toxic substance concentration of the carnivorous insects and he may consider the toxic substance concentrations in plants, herbivorous insects and soil to be of less importance. He may, therefore, choose a weight of 10 for the first state variable and only 1 or 2 for the subsequent three.

If it is impossible to calibrate a model properly, it is not necessarily due to an incorrect model, but may be due to poor quality of the data. The quality of the data is crucial for calibration. Furthermore, it is of great importance that the observations reflect the real dynamics of the system. If the objective of the model is to give a good description of one or a few state variables, it is, of course, particularly essential that the data can show the dynamics of just these internal variables. The frequency of the data collection should therefore reflect the dynamics of the focal state variables. This rule has unfortunately often been violated in modelling.

It is strongly recommended that the dynamics of all state variables are considered before the data collection program is determined in detail. Frequently, some state variables have particularly pronounced dynamics in specific periods – often in spring – and it may be of great advantage to have a dense data collection in this period, in particular. In [10] it is shown how a dense data collection program in a certain period can be applied to provide additional certainty for the determination of some important parameters. This question will be discussed further in Chapter 3.

From these considerations, recommendations can now be drawn about the feasibility of carrying out a calibration of a model in ecology:

1. Find as many parameters as possible from the literature (see Fig. 1.8) [8, 9]. Even a wide range for the parameters should be considered very valuable, as approximate initial guesses for all parameters are urgently needed.
2. If some parameters cannot be found in the literature, which is often the case, estimation methods should be used (see also Chapter 5). For crucial parameters, it may be better to determine them by experiments *in situ* or in the laboratory.
3. A sensitivity analysis should be carried out to determine which parameters are most important to be known with high certainty.
4. The use of an intensive data collection program for the most important state variables should be considered to provide a better estimation for the most crucial parameters. It is generally not possible to uncover the dynamics of state variables by observations with too little frequency that do not reflect the dynamics of the system. For instance if a model considers the diurnal variation of photosynthesis and respiration for plants, or the diurnal variations are of utmost importance for the core problem of the model, it is absolutely necessary to build the model on the basis of the observations made several times per 24 hours. For further details about the method, see the description in Chapter 3.
5. First, at this stage, the calibration should be carried out by use of the data not yet applied. The most important parameters are selected and the calibration is limited to these, or, at the most, to 5–10 parameters. In the first instance, the calibration is carried out by using the trial-and-error method to get acquainted with the model reaction to changes in the parameters. An automatic calibration procedure may be used afterward to polish the parameter estimation.
6. These results are used in a second sensitivity analysis, which may give results different from the first sensitivity analysis.
7. A second calibration is now used on the parameters that are most important according to the second sensitivity analysis. In this case too, both the above-mentioned calibration methods may be used. After this final calibration, the model can be considered calibrated and we can go to the next step: validation.

The calibration should always be followed by a validation. In this step, the modeller tests the model against an independent set of data to observe how well the model simulations fit these data or,

Figure 1.8: One of the major literature sources for parameters (see [9]).

expressed differently, to determine the uncertainty of the model. It must, however, be emphasized that the validation only confirms the model behavior under the range of conditions represented by the available data. Therefore, it is preferable to validate the model using data obtained from a period in which other conditions prevail. For instance, when a eutrophication model is tested, it should preferably have data sets for the calibration and the validation, which differ by the level of eutrophication. If an ideal validation cannot be obtained, it is, however, still of importance to validate the model. The method of validation is dependent of the objectives of the model. A comparison between measured and computed data for calculation of the standard deviation between model results and observations is often applied. This is, however, often not sufficient, as it may not focus on the main objectives of the model, but only on the general ability of the model to describe correctly the state variables of the ecosystem. It is necessary, therefore, to translate the main objectives of the model into a few validation criteria. They cannot be formulated generally, but are individual for the model and the modeller. For instance, if we are concerned with the eutrophication in an aquatic ecosystem, it would be useful to compare the measured and computed maximum concentrations of phytoplankton, because the minimum water quality is often the core concern. The discussion of the validation can be summarized by the following issues:

1. Validation is always required to get a picture of the reliability of the model or, expressed differently, to determine the model uncertainty.
2. Attempts should be made to get good data for the validation, which are entirely different from those used in the calibration. Furthermore, it is important to have data from a wide range of forcing functions that are defined by the objectives of the model. The uncertainty determined by the validation can be used directly to assess the expected uncertainty of prognoses, provided, of course, that the validation was carried out with forcing functions as close as possible to the forcing functions that are used in the prognoses. In this context, it is extremely important that the modeller makes it completely clear for the users of the model that all prognoses inevitably must have uncertainty and that the use of the prognoses must be based on a good knowledge of the uncertainty, which means that a good validation of the model is compulsory.
3. The validation criteria are formulated based on the objectives of the model and the quality of the available data. The main purpose of the model may, however, be an exploratory analysis to understand how the system responds to the dominating forcing functions. In this case, a structural validation is probably sufficient, implying that the changes can be explained qualitatively.

1.5 Selection of model complexity

It is clear from the discussion of the modelling procedure that in our model we should, on one hand, include all the important (for solution of the problem) state variables, processes and forcing functions, but, on the other hand, not make the model unnecessarily complex, because that would inevitably imply that it would contain more parameters that have to be determined in one way or another. The modeller is therefore left with the obvious questions: How complex should I build the model? And is there a method to select the complexity?

The answer to the second question is yes, by "trial and error". It is, of course, possible to examine how much further additions to the complexity would change the results, by testing these additions and observing how much the crucial state variables are changed. It is also possible to find the sensitivity of the additional parameters resulting from the additional complexity. From such examinations, it would be possible to see if additional complexity would increase or reduce the uncertainty of the model results. It would also be possible to go one step further, namely to validate different editions of the model with different complexity. Such comprehensive examinations are possible but, of course, also cumbersome.

It would therefore be beneficial to have some simple rules to ensure at least an approximate right choice of the complexity in the first instance. Costanza and Sklar [11] have examined 88 different eutrophication models and the results are shown in Fig. 1.9. The complexity = the articulation = the number of state variables + 0.1 × the number of boxes + 0.01 × the number of time steps. The effectiveness is the articulation or complexity divided by the uncertainty.

From this examination, it is clear that an intermediate complexity is preferable. The right complexity is, so to say, a tradeoff between knowing nothing about everything (the model is too

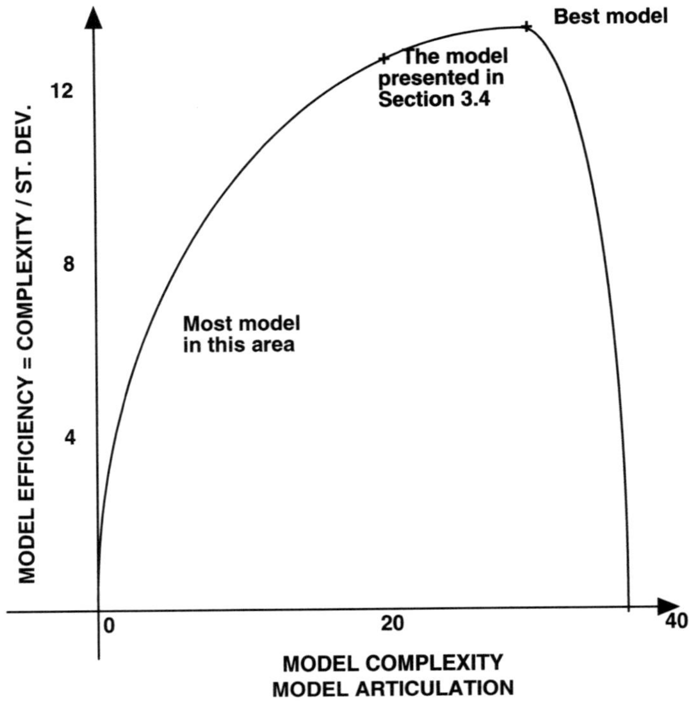

Figure 1.9: The graph shows the results of an examination of 88 eutrophication models. The complexity = the articulation = the number of state variables + 0.1 × the number of boxes + 0.01 × the number of time steps. The effectiveness is the complexity divided by the uncertainty.

complex and therefore too uncertain) and everything about nothing (the model complexity is too low and the focal problem cannot be solved by the model).

From the graph, it can be seen that a complexity (see definition above) of about 20 (8–35) seems to be recommendable. This rule should, however, be used very carefully, because the right complexity will always be dependent on the required accuracy for the focal state variables, the possibilities to simplify from the real ecosystem to the model, our additional knowledge about the parameters and the processes, and the costs of the data collection.

Another rule of thumb for the selection of the complexity should be mentioned in this context: The absolute maximum number of parameters to be calibrated should be the square root of the number of available observations. It means that a model with 10 parameters to be calibrated needs 100 observations. If the number of state variables is 5 too, 20 sets of observations of these 5 state variables are needed (20 sets of 5 observable state variables each). The 10 parameters can, in this context, be considered unknown and with 100 observations we have 100 equations to determine the 10 parameters, which gives 10 observations to determine each parameter. Ten observations to determine a very important number (an important parameter in this case), should be considered as minimum. It would of course be beneficial to have more observations for the determinations of the parameters than the 100 indicated in this case.

Moreover, it should be repeated that it is, of course, very beneficial to get as much knowledge as possible about all the applied parameters before the calibration. Therefore, it is recommendable to make a comprehensive literature search about the parameters and use all available parameter estimation methods before the calibration is carried out. The three most applied estimation methods are:

1. the use of mass and energy balances to estimate which parameter range is feasible,
2. the use of allometric principles and
3. the use of very frequent measurements in selected periods.

All three methods are presumed to be applied whenever it has been mentioned, in the previous sections, that parameter estimation methods should be considered. The methods will be presented in Chapters 3 and 6, when examples are at hand where these methods would demonstrate a clear advantage for the estimation of crucial parameters.

Finally, it should be emphasized that the better the modeller knows the problem and the system, the better he is able to choose the right complexity and that modelling will always contain an element of intuition.

1.6 Model uncertainty

Validation is a very important modelling step, because it will give the uncertainty of the model results. It attempts to answer the question: How to account for the model uncertainty in the use of the model in environmental management, for instance, by development of prognoses for different environmental strategies? If we use the model as a research tool, the validation will tell us whether the model results can be used to support or reject a hypothesis. The uncertainty determined by the validation relative to the difference between the hypothesis and the model results will be decisive.

The validation result can, of course, also be used to consider which model revisions we could apply to reduce the uncertainty. In our effort to improve the model, we should ask the following pertinent questions:

1. What is the uncertainty of the observations (measurements)? If the uncertainty of the model is not very different from the uncertainty of the observations, it will probably be beneficial to get more reliable observations with less uncertainty.
2. Are the observations representing the dynamic of the system? If not, it should be considered to have more frequent measurements for some period or revise the measuring program to be able to better capture the dynamic of the system. See the discussion of this question in Chapter 3.
3. Are some important processes or components missing in the model? This is, of course, a repetition of the discussion about the model complexity. As mentioned previously, in this context it is important to set up a mass and/or energy balance to reveal the important processes and sources.
4. Is the model giving wrong descriptions of important processes? Should we, in other words, change some of the process equations? This question was also asked when we developed the model, but it would be natural to raise the question again, if the model uncertainty is not satisfactory.
5. It can be recommended to give a sufficiently comprehensive answer to question 4 and use the model experimentally to find the best answer. It is quite easy, in most cases, to replace important equations by other expressions or add new components or processes and so on. Such experiments are very important to elucidate the process formulations. Small changes in process equations that make big changes of the model results uncover the soft points of the model and may inspire additional experiments or observations *in situ* or in the laboratory or further changes of the model.

These five points have all been discussed previously in the chapter, but in another context. However, it is, important to repeat these five points when we are discussing the possibilities to improve the model, particularly for a reduction of the model uncertainty. It should also be emphasized that the "ideal" model can never be achieved, but by steadily questioning the model and using the five points again and again, we can asymptotically approach the ideal model step by step. However, the ideal model is not necessarily to have a useful and powerful tool in environmental management and ecosystem research. A satisfactory calibration and validation with sufficiently low uncertainties to allow application in a defined context would be the general requirement for the pragmatic modeller.

1.7 Mediated modelling

The following questions arise as a result of the proposed model procedure (see Fig. 1.3):

- How is it possible to consider all the different aspects of the problem and its solution, when a very wide spectrum of expertise is represented in the modelling team?
- How is it possible to consider all relevant ecosystem properties?
- How is it possible to integrate insights from many different disciplines as, for instance, ecology, economy and social science?
- How can we ensure that all important stakeholders are included in the process?
- How is it possible to integrate impacts at different scales?
- How can the policymakers and the stakeholders see the consequences of their action plan over longer time scales?
- How is it possible to understand the very root of the problems and their sources?
- How is it possible to build the best environmental consensus?

Institutionalized or mediated modelling (IMM) can give the answer to these questions. The IMM is described in more detail in the book by Jørgensen, *et al.* [12]. The main idea is to represent all stakeholders, policymakers, managers and scientists that have knowledge and ideas about the problem, the system and the possible solutions in the modelling procedure. The model becomes simultaneously the common language among members of a multidisciplinary team.

The model is developed as a result of an integrated brainstorming meeting. For most model development, depending on the complexity of the problem and the system, several days are required to reach a satisfactory model that can be used as a tool to solve the problem.

The advantages of IMM are that (partly taken from [13]):

1. the level of shared understanding increases;
2. a consensus is built about the structure of a complex problem for a complex system, because all interests are represented in the stepwise model development;
3. the result of the modelling process and the model serves as a tool to disseminate the insights gained by the modelling procedure;
4. the effectiveness of the decision making is increased;
5. team building is facilitated;
6. the process is emphasized over the product;
7. the state-of-the-art knowledge is captured and organized.

Brown [14] evaluates the level of "groupiness" based on the following five criteria:

1. The individual members perceive that they are a part of the group.
2. A group is oriented toward a common goal.

3. Interaction between group members exists.
4. A realization of interdependence is present.
5. There is a structure of roles/status and norms.

For further details about IMM, see [12, 13].

Illustration 1.1

To illustrate how to go about modelling, we will use the conceptual model in Fig. 1.4. Let us presume that the conceptual diagram, showing the state variables and the forcing functions, is in accordance with the available observations. It should be discussed whether we have omitted important state variables, for instance, omnivorous zooplankton species. If our knowledge of the ecosystem is sufficiently detailed to ensure that omnivorous zooplankton species are insignificant, it would, of course, be all right not to include them. However, if we, still have some doubt about the importance of omnivorous zooplankton, it is possible, of course, to include them and note how much it would change the results – particularly the phytoplankton concentration, as it is a eutrophication model where phytoplankton will be considered the most important state variable.

Another question could be: Have we omitted important processes? To answer this question, we need to go through a list of all possible processes. The model has 7 state variables, which means that there are 42 possible connections. It is recommended to ask for each of these 42 connections: Is this connection possible in nature? Often an adjacency matrix, as shown in Fig. 1.10, is applied as a checklist. If a connection is realized, it is indicated with "1" and if it is not realized with "0".

Solution

We can illustrate the selection of equations by considering, for instance, the state variable ammonium. The differential equation is:

$$\text{dAmmonium/dt} = \text{Process 12} + \text{Process 9} + \text{Process 13} - \text{Process 1} - \text{Process 14} + \text{ammonium in (16)} - \text{ammonium out (16)}.$$

From: \ To:	Ammonium	Nitrate	Phytoplankton	Zooplankton	Detritus	Fish	Sediment
Ammonium	–	1	1	0	0	0	0
Nitrate	0	–	1	0	0	0	0
Phytoplankton	0	0	–	1	1	0	1
Zooplankton	1	0	0	–	1	1	0
Detritus	1	0	0	0	–	0	1
Fish	0	0	0	0	1	–	0
Sediment	1	0	0	0	0	0	–

Figure 1.10: Adjacency matrix for the conceptual diagram in Fig. 1.4. Twelve out of the 42 possible connections are realized. It means that the ecological network presented in the conceptual diagram has a connectivity of 12/42 = 0.29. The diagram indicates 22 processes. Eight of these processes are forcing functions. The processes 5 and 20 and the processes 7 and 21 are combined to make two processes only in the matrix. Furthermore, the uptake of ammonium and nitrate by phytoplankton can be described as one process in the system of equations.

Process 12 (excretion of ammonia by zooplankton) may be expressed as a first order reaction, which means that Process 12 = Zooplankton-N $\times$ K_{ex}.

Process 9 may be expressed as a first order reaction, which means that Process 9 = Detritus-N $\times$ K_1.

Process 13 may be expressed as a first order reaction, which means that Process 13 = Sediment-N $\times$ K_2.

Process 1 is usually expressed as a Michaelis–Menten equation, which means that Process 1 = U_A $\times$ [Phytoplankton-N] $\times$ [Ammonium]/([Ammonium] + k_u) and a similar expression for the nitrate uptake.

Process 14 is usually expressed as a Michaelis–Menten equation, which means that Process 14 = K_N $\times$ [Ammonium]^2/([Ammonium] + k_m) correspondingly for the uptake of nitrate.

Ammonium-in = Q $\times$ [Ammonium-in], where Q and [Ammonium-in] are forcing functions, known as table functions (function of time) or by use of equations.

Ammonium-out = Q $\times$ [Ammonium-out], where we presume that Q-in = Q-out – no or minor accumulation of water takes place in the aquatic ecosystem. It is correct with good approximations in many cases, but not in all cases. It also means that the precipitation and evaporation are equal, which in many cases, but not in all cases, is correct with a good approximation. Usually [Ammonium] can be used to replace [Ammonium-out].

The following parameters should be determined (a literature range is indicated in brackets after the parameter definition and unit):

K_{ex}: ammonium excretion coefficient for zooplankton, 1/24 h (0.001–0.01);
K_1: mineralization rate coefficient, 1/24 h (0.001–0.05);
K_2: sediment release rate coefficient, 1/24 h (0.002–0.05);
U_A: uptake rate coefficient by phytoplankton for ammonium, 1/24 h (0.5–5);
k_u: Michaelis–Menten half-saturation constant, mg/l (0.02–0.2);
K_N: nitrification rate coefficient, 1/24 h (0.005–0.1);
k_m: Michaelis–Menten half-saturation constant, mg/l (0.01–0.5).

Determination of the 7 parameters by calibration would require at least $7^2 = 49$ observations. It means seven sets of observations of the seven state variables. If the model has 4 boxes and is used for 1825 days or 5 years (one time step = 1 day), the complexity is 7 + 0.4 + 1.8 = 9.2. It means that the complexity is on the low side. It cannot be excluded that the model beneficially could be expanded slightly, for instance, to include 1–3 more state variables.

References

[1] Jørgensen, S.E. & Bendoricchio, G., *Fundamentals of Ecological Modelling*, 3rd edn, Elsevier: Amsterdam, 2001.

[2] Chen, C.W., Concepts and utilities of ecological models. *Proc. Am. Soc. Civ. Eng. J. San. Eng. Div.*, **96(SA5)**, pp. 1085–1097, 1970.

[3] Christensen, V., On ECOPATH. Fishbyte and fisheries management. *Fishbyte*, **9(2)**, pp. 62–66, 1991.

[4] Christensen, V., *Network Analysis of Trophic Interactions in Aquatic Ecosystems*, PhD Diss., Copenhagen University, 1992.

[5] Jørgensen, S.E., *Integration of Ecosystem Theories: A Pattern*, 3rd edn, Kluwer Acad. Publ.: Dordrecht, Boston, London, 2002.

[6] Jørgensen, S.E. & Svirezhev, Y., *Toward a Thermodynamic Theory for Ecological Systems*, Elsevier: Amsterdam, Oxford, 2005.

[7] Jørgensen, S.E. & Fath, B., *A New Ecology. A Systems Approach*, Elsevier: Amsterdam, Oxford, 2007.

[8] Jørgensen, S.E., Jørgensen, L.A. & Nielsen, S.N., *Handbook of Environmental Data and Ecological Parameters*, 2nd edn, Elsevier: Amsterdam, 1991.

[9] Jørgensen, L.A., Jørgensen, S.E. & Nielsen, S.N., *Ecotox*, CD-ROM, Elsevier: Amsterdam, 2000.

[10] Jørgensen, S.E., Parameter estimation in eutrophication models. *Ecol. Modelling*, **13**, pp. 111–129, 1981.

[11] Costanza, R. & Sklar, F.H., Articulation, accuracy and effectiveness of mathematical models: a review of freshwater wetland applications. *Ecol. Modelling*, **27**, pp. 45–69, 1985.

[12] Jørgensen, S.E., Chon, T. & Recknagel, F., *Handbook of Ecological Modelling and Informatics*, WIT Press: Southampton, 2009.

[13] Van den Belt, M., *Mediated Modeling*, Island Press: Washington, 2004.

[14] Brown, R., *Group Processes: Dynamics Within and Between Groups*, Blackwell: Oxford, 1988.

Exercises/problems – Chapter 1

1. Explain the importance of verification, calibration and validation. Can models without these three steps be developed at all?
2. How often would you determine the phytoplankton concentration, if a model for the diurnal variations of primary production was supposed to be determined?
3. Draw a conceptual diagram of an ecological model considering the oxygen depletion in a pond due to discharge of organic matter. The oxygen production due to photosynthesis is very minor and need not to be considered. Decomposition of organic matter and re-aeration are the main processes that determine the oxygen concentration. The pond has no in- or outflows. The precipitation (in mm/day) is known, the pond is 5 meters deep and the temperature as a function of time is known too. The rainwater is presumed to be saturated by air. List the state variables and the forcing functions. Give the differential equations and the process equations and list the parameters. How could the contribution from photosynthesis be included in the model?
4. Discuss the complexity of the model developed in Exercise 3. How many observations would you recommend to have an acceptable calibration?
5. The figure below illustrates a conceptual diagram for a model.

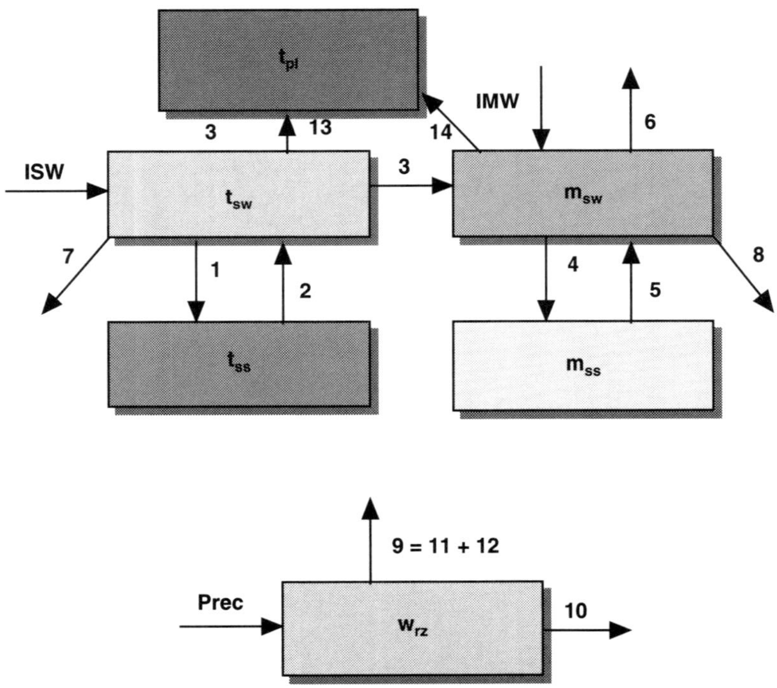

The model has six state variables: t_{sw}, tylosine (an antibiotic used for pigs) in soil water; t_{ss}, tylosine in soil (adsorbed), m_{sw}, metabolites of tylosine in soil water; m_{ss}, metabolites in soil (adsorbed), t_{pl}, tylosine in plants and w_{rz}, water in the root zone. ISW is a forcing function that accounts for how much tylosine is added with manure and IMW is the amount of metabolites contained in the manure. Processes 7 and 8 are forcing functions determining the amount of tylosine and metabolites lost to ground water; 6 is the loss of metabolites by decomposition. Prec is the forcing function giving the precipitation, while 10, 11 and 12 are the forcing functions giving information about the run-off and the evaporation from soil and plants, respectively.

Develop an adjacency matrix for the diagram; discuss the complexity of the model and the number of observations needed for the calibration, presuming two parameters per process. Propose two to four additional state variables to cover the fate of tylosine in more detail using a slightly more complex model.

CHAPTER 2

Model types

2.1 Classification of models

In Chapter 1, a modelling procedure is presented and it is discussed how it is possible to select close to optimum complexity. In the 1970s, when ecological models started massively to be applied in environmental management and as a scientific tool in systems ecology, most of the applied models belonged to three types: population dynamic models, bioenergetic models and biogeochemical models. For the first type, conservations of the number of individuals in a population were applied to set up the differential equations. For the second and third type, the conservation of energy and/or mass was the key principle applied for the development of the differential equations; see Table 2.1. For all the three types, there is a dynamic version using the differential equations and a steady state version using steady state, which means that the differential equations are all equal to zero. The steady state versions are used, when the problem can be solved by presentation of an average situation or a worse-case situation, presuming steady state in both the situations. In data-poor situations it is often beneficial to apply the steady state versions, because the quality and the quantity of the data are not sufficient to develop a dynamic model. All the three types can be developed by following the procedure in Chapter 1. The equations are also developed similarly for the three model types. The discussion and the illustration presented in Illustration 1.1 are valid for all the three model types.

It was very obvious which type and version is to be applied in which cases, because the problem defined generally the model type. In the last 2–3 decades, several new types of models have, however, emerged to cover a wide spectrum of more specific problems. It makes, of course, the selection of model type more cumbersome. It is therefore crucial to have a good overview of the available model types and their characteristics to be able, already in the first step of the model procedure, to choose the best model type that is able to meet the model objectives. It is the focus of this chapter.

2.2 Model types – an overview

The new model types, developed during the last couple of decades, have been developed to answer a number of relevant modelling questions, which arose as a result of the increasing use of ecological models in the 1970s. The relevant modelling questions are listed below:

1. How can we describe the spatial distribution that is often crucial to understand ecosystem reactions and to select the best environmental strategy?
2. Ecosystems are middle number systems [1] in the sense that the number of components is smaller in magnitude than the number of atoms in a system All the components are different

Table 2.1: Classification of models.

Type of models	Organization	Pattern	Measurements
Biodemographic	Conservation of number of individuals	Life cycles of species	Number of species or individuals
Bioenergetic	Conservation of energy	Energy flow	Energy
Biogeochemical	Conservation of mass	Element cycles	Mass or concen-tions

and that is often important for a proper description of the ecosystem reactions to consider the differences in properties among individuals.

3. The species are adaptable and may change their properties to meet the changes in the prevailing conditions – it means the forcing functions. Furthermore, the species may, by more drastic changes, be replaced by other species better fitted to the combinations of forcing functions. How to account for these changes? Even the networks may be changed if more biological components with very different properties are replaced by other species. How to account for these structural changes?
4. Can we model a system that has only relatively uncertain observations/data?
5. The forcing functions and several ecological processes are stochastic. How to account for that?
6. How to develop models if the database is very heterogeneous, that is, based on observations from many different ecosystems?
7. How to develop models of ecosystems, when our knowledge is mainly based on a number of rules/properties/propositions?
8. How to consider toxic substances in the environment? Does development of a toxic substance model require a special model type?
9. How to consider a dynamic transport in an ecosystem, even sometimes in three dimensions (3D), included descriptions of gradients, also sometimes in 3D?
10. How to consider several linked ecosystems simultaneously? Can we model entire landscapes consisting of several ecosystems?

These questions have all found an answer today, mainly by the development of new model types that are able to cope with the defined problems. Spatial models often based on the use of geographical information system (GIS) have been developed to be able to come with an answer to question 1. Individual based models (IBMs) are able to give an answer to question 2. The software that can be used to develop IBMs is even available to facilitate the development of IBMs. The software can also be utilized to cover spatial distribution (see question 1). Structurally dynamic models (SDMs) have been developed to solve the problem expressed in question 3. Fuzzy models can be used to make models based on a poor and may be only a semi-quantitative database. Stochastic models are applied today in ecological modelling, although the application is still not wide, probably because it is not very frequent that there is an urgent need to include stochastic processes in ecological models. Artificial neural networks (ANNs) is the answer to question 6. IBMs can, in many cases, meet the demands expressed in question 7. Ecotoxicological models have been widely used, particularly in the last 10–15 years, because the need is very obvious.

Lately, there has been an increasing use of a coupling between 3D models for instance, hydrological 3D models and ecological models. It has also opened for the development of landscape

Table 2.2: Application of the most general model types for the period 1975–1980 with the period 2001–2006.

Model type	Percentage application 1975–1980	Percentage application 2001–2006
Dynamic biogeochemical models	62.5	32.0
Steady state biogeochemical models	0	1.8
Population dynamic models	31.0	24.9
Spatial models	0	19.1
SDMs	1.5	8.0
IBMs and cellular automata	0	5.2
ANN and use of artificial intelligence	0	4.9
Fuzzy models	0.5	1.8
Ecotoxicological models	0	2.2

models. The power of today's computers has, of course, facilitated the general use of more and more complex models, including 3D models and relatively large landscape models.

This book presents the development of biogeochemical models, population dynamic models, spatial models, ecotoxicological models and SDMs. These five types are the most used types (see for instance the model statistics in Jørgensen *et al.* [2]). For all the types of models, a very comprehensive overview with examples of all the above mentioned model types on an enclosed CD can be found in this handbook [2]. It is therefore recommendable to supply this textbook of modelling with a class copy of the *Handbook of Ecological Modelling and Informatics* [2].

Table 2.2 gives a summary of a model statistics based on the number of publications in the journal *Ecological Modelling*. The percentage application of the most general model types are compared for the period 1975–1980 with the period 2001–2006. In the table, we have distinguished between nine types:

1. dynamic biogeochemical models;
2. steady state biogeochemical models;
3. population dynamic models;
4. spatial models;
5. structurally dynamic models, SDM;
6. individual-based models and cellular automata;
7. ANN
8. fuzzy models and
9. ecotoxicological models.

The data in Table 2.2 is shown graphically in Figs. 2.1–2.2.

The number of papers published during the period 2001–2006 is, however, about nine times the number of papers published during the period 1975–1980. It means that the number of dynamic biogeochemical model papers published recently is more than 4.5 times the number published during the late 1970s and that the number of papers on structurally dynamic modelling has increased almost by factor 50 during the last 35 years. A comparison of the Figs 2.1 and 2.2 also shows that the spectrum of model types applied today is much wider than about 30 years ago. It is, of course, not surprising as the new types of models have been developed because there was an urgent need to answer the 10 modelling questions listed above.

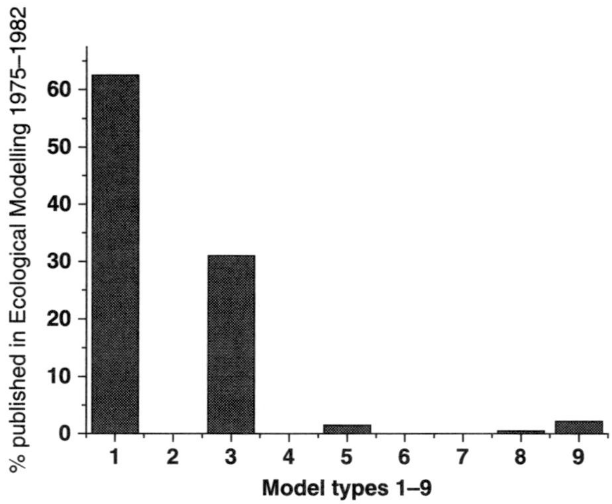

Figure 2.1: The percentage of papers published in *Ecological Modelling* of the nine model types 1975–1982 listed above and in Table 2.1.

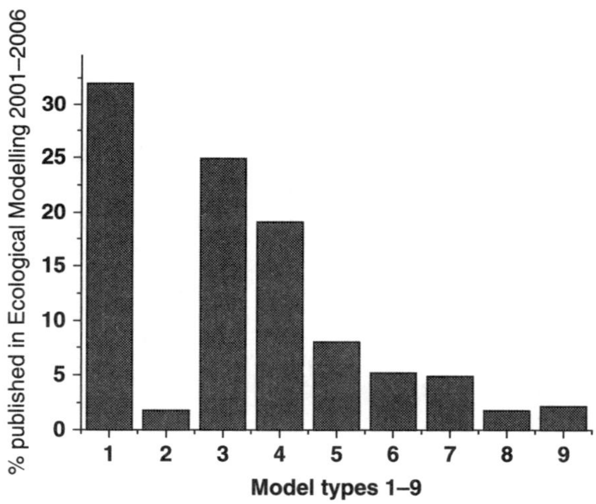

Figure 2.2: The percentage of papers published in *Ecological Modelling* of the nine model types 2001–2006 listed above and in Table 2.1.

With the present spectrum of model types it is possible to solve all the major modelling problems that started to be raised already in the 1970s (see the above presented 10 questions), which, of course, has increased the application of ecological models in general and, of course, particularly for the new model types. However, it is also clear that all the problems cannot be solved completely. We still have a number of problems that may not be possible to solve by the use of a single model type. The very complicated problem requires may be the use of hybrid models, that is, by a combination of the above presented model types.

New model types may be developed during the coming decade to solve the complicated problems, which today would require the use of hybrid models. It is, however, agreed among ecological modellers that we currently have sufficient model types to solve further most of ecological modelling problems being faced with us nowadays, although some modelling case studies still need to guarantee their feasibility of all model developments in real situations.

2.3 Characteristics of the model types

For all the main model types available today, the characteristics of the model type, the advantages, the disadvantages (mostly expressed as a limitation of the application) will be given below one by one. The applicability of the various model types will be discussed in the next section. The application of catastrophe theory and chaos theory are not included in the overview below, because they are considered mathematical tools that in principle can be applied as useful tools in the development of several different model types. Furthermore, statistical models will also not be mentioned because statistics is considered as a tool that can be applied in ecological modelling to give a better process description. If a model is based entirely on the application of statistics, it is denoted as black box models, because it has no causality, but based entirely on statistics. Black box models can hardly be used to uncover new ecological knowledge. They may be used as management tool, but as they are not based on ecology but on statistics, they can hardly be denoted as ecological models. As ecological modelling is a strong tool in the science ecology, where the focus to a high extent is on causality, the use of black or dark grey box models is avoided to the extent possible.

A short summary of the most generally applied models are given below. As previously mentioned, a more comprehensive treatment can be found in Jørgensen *et al.* [2].

2.3.1 Dynamic biogeochemical and bioenergetic models

This model type is widely used, as can be seen in Table 2.2. The model type applies generally differential equations to express the dynamics. Change in state variables are expressed as the results of the ingoing minus the outgoing processes and the model is, therefore, based on the conservation principles. The process equations are based most commonly on causality; but a submodel can in principle also be a result of a statistic analysis of observations. The model type has some clear advantages that make it attractive still to use this model type for the development of many models.

The advantages are:

- Most often based on causality.
- Based on mass or energy conservation principles.
- Easy to understand, interpret and develop.
- Software is available, for instance STELLA.
- Easy to use for predictions.

The disadvantages are:

- Can hardly be used for heterogeneous data.
- A relatively good database is required.
- Difficult to calibrate when they are complex and therefore contain many parameters.
- Can hardly account for adaptation and changes in species composition.

The advantages and disadvantages define, so to say, the area of application: for description of the state of an ecosystem, when a good data set is available. The model type has been extensively applied in environmental management as a powerful tool to understand the reactions of ecosystems to pollutants and to set up prognoses.

2.3.2 Steady state biogeochemical models

Due to the limitations of this model type, it has not been used in more than 1.8% of the publications in *Ecological Modelling* during the period 2001–2006. The model type is a biogeochemical or bioenergetic dynamic model where the differential equations are all set to zero to obtain the values of the state variables corresponding to the static situation.

The advantages are:

- Require generally smaller databases than other types.
- Are excellent to give a worse case or average situation.
- The results are easily validated (and verified).

The disadvantages are:

- Does not give any information about dynamics and changes over time.
- Prediction with time as independent variable is not possible.
- Can only give average or worse case situations.

This model type will often be used when a static situation is sufficient to give a proper description of an ecological system or to take environmental management decisions.

2.3.3 Population dynamic models

This model type is rooted in the Lotka–Volterra Model that was developed in the 1920s. Numerous papers have been published about the mathematics behind this model and a number of deviated models. The mathematics of these equation systems is not very interesting from an ecological modelling point of view, where the focus is a realistic description of ecological populations. Population dynamic models may include age structure, which in most cases is based on matrix calculations. The number of population dynamic papers is five times as much today as in the late 1970s, which illustrates that ecological modelling has developed significantly from the 1970s to today. The minor reduction in percentage is due to the application of a much wider spectrum of different model types today.

The advantages are:

- Able to follow the development of a population.
- Age structure and impact factors can easily be considered.
- Easy to understand, interpret and develop.
- Most often based on causality.

The disadvantages are:

- The conservation principles are sometimes not applied, although it is easy in most cases.
- Limited to population dynamics.
- Require, as for biogeochemical dynamic models, a relatively good and homogenous database.
- Difficult in some situations to calibrate.

This model type is typically applied to keep a track on the development of a population. Number of individual is the most applied unit, but it can of course easily be translated into biomass or bioenergy. Effects of toxic substances on the development of populations can easily be covered by increasing the mortality and decreasing the growth corresponding to the effect of the toxic substance. The model type is extensively used in management of fishery and other natural resources and national parks.

2.3.4 Structurally dynamic models

This type of models changes the parameters, corresponding to the properties of the biological modelling components, to account for adaptation and changes in species composition. It is possible either to use knowledge or artificial intelligence to describe the changes in the parameters. Most often, however, it is used as a goal function to find the changes of the parameters. The thermodynamic variable, eco-exergy has most often been used as goal functions in SDMs. Minor changes of the parameters may be due to adaptation to the changed conditions, but for major changes it is most probably a change in the state variables – a shift in the species composition – that is causing the change in the parameters. It may also be possible to use the approach to a major change in the ecological network, although no reference to this application of the structurally dynamic modelling approach is yet available.

SDMs are applied much more today than 25–30 years ago.

The model type has the following advantages:

- Is able to account for adaptation.
- Is able to account for shift in species composition.
- Can be used to model biodiversity and ecological niches.
- The parameters determined by the goal functions do not need to be calibrated.
- Relatively easy to develop and interpret.

The disadvantages of this model type are:

- Selection of a goal function or use of artificial intelligence is needed.
- Usually computer time consuming.
- Require information about structural changes for the calibration and validation.
- No available software; programming needed.

This model type should be applied whenever it is known that structural changes take place. It is also recommended for models that are used in environmental management to make prognoses resulting from major changes in the forcing functions (impacts).

2.3.5 Fuzzy models

Fuzzy models may either be based on knowledge (it is called the Mamdani type) or based on data (it is called the Sugeno type). The Mamdani type models are based on a set of linguistic expert formulations and it is applied when no data are available. The Sugeno type applies an optimization procedure and it is applied when only uncertain data are available.

This model type has the following advantages:

- Can be applied on a fuzzy data set.
- Can be applied on semi-quantitative (linguistic formulations) information.

- Can be applied for the development of models, where a semi-quantitative assessment is sufficient.

This model type has the following disadvantages:

- Can hardly be used for more complex model formulations.
- Cannot be used where numeric indications are needed.
- No software available to run this type of models, although there are facilities in Mat. Lab to run fuzzy models.

The application of this model type is obvious. It is applied when the data set is fuzzy or only semi-quantitative expert knowledge is available, provided, of course, that the semi-quantitative results are sufficient for the ecological description or the environmental management.

2.3.6 Artificial neural networks

This type of models is able to give relationships between state variables and forcing functions based on a heterogeneous database. It is in principle a black box model and is, therefore, not based on causality; but it gives in, most cases, very useful models that can be applied for prognoses, provided that the model has been based on a sufficient big database that allows to find the relationships and to test it afterwards on an independent data set. This model type was not applied before 1982 in ecological modelling. The model type can be developed by the use of available software or by the use of Mat. Lab. ANN including self-organizing maps and the use of artificial intelligence in modelling have the following advantages:

- May be used where other methods must give up.
- Easy to apply.
- Give a good indication of the certainty due to the application of a test set.
- Can be used on a heterogeneous data set.
- Give a close to optimum use of the data set.

The disadvantages can be summarized in the following points:

- No causality, unless algorithms are introduced or a hybrid between ANN and another model type is applied.
- Cannot replace biogeochemical models based on the conservation principles.
- The accuracy of predictions sometimes limited, although validation almost always is used.

The advantages and disadvantages of this model type indicate where it would be advantageous to apply ANN, namely where ecological descriptions and understandings are required on basis of a heterogeneous database, for instance, data from several different ecosystems of the same type. It is also often applied beneficially when the database is more homogenous, for instance, when the focus is on a specific ecosystem, although the modeller should seriously consider using biogeochemical dynamic models due to their causality. ANN is, however, faster to use and the time consuming calibration that is needed for biogeochemical models is not needed.

2.3.7 Spatial models

As the individual differences may be crucial for the model results, the spatial differences of the forcing functions, of the non-biological state variables and of the biological state variables may be decisive for the model results, too. Furthermore, it may be required to obtain model results that reveal the spatial differences, because they may be needed to understand the ecological reactions or to make a proper environmental management. Models that give the spatial differences must, of course, also consider the spatial differences in the processes, forcing functions and state variables. Due to the urgent needs for a proper description of the spatial differences, it is not surprising that *Ecological Modelling* has published almost 250 papers about spatial modelling from 2001 to 2006 and that the number of models that focus on spatial distribution is increasing rapidly.

There are a number of possibilities to cover the spatial differences in the development of an ecological model. It is not possible to cover all of them here; but as mentioned under IBMs, cellular automata may be used in this context. GIS is another possible approach to be applied for the development of what is denoted as surface models. GIS is also a convenient method to present the results of all kinds of spatial models. For aquatic ecosystems, the ultimate spatial model would give a 3D description of the processes, the forcing functions and the state variables. It is often a question about a good description of the hydrodynamics. There has been an increasing use of models that couple 3D hydrodynamic models and ecological models.

Spatial models offer the following advantages:

- Cover spatial distribution that often may be of importance in ecology.
- The results can be presented in many informative ways, for instance, GIS.

The disadvantages are:

- Require usually a huge database giving information about the spatial distribution.
- Calibration and validation are difficult and time consuming
- Sometimes a very complex model is usually needed to give a proper description of the spatial patterns.

The spatial models are applied whenever it is required that the results include the spatial distribution, because it is decisive or the spatial distribution is crucial for the model results.

2.3.8 Individual based models and cellular automata

This model type can be regarded as a reductionistic approach, deriving the properties of a system from the properties and interactions among elements of the system. The model type was developed because all the biological components in ecosystems have different properties, which is not considered in generally applied biogeochemical models. Within the same species the differences are minor and are therefore often neglected in biogeochemical models, but the differences among individuals of the same species may sometimes be important for the ecological reactions. For instance, individuals may have different size, which gives different combinations of properties as it is known from the allometric principles (see Chapter 6). The right property may, of course, be decisive for growth and/or survival in certain situations, as it is recognized by all modellers. Consequently, a model without the differences among individual may give a completely wrong result.

Cellular automata are systems of cells interacting in a simple way but displaying complex overall behavior. They are usually characterized by a few salient features. Cellular automata form a class of spatio-dynamical models where time, space and states are discrete. IBMs are often using the cellular automata approach, although there are IBMs that are not cellular automata models. Furthermore, there are cellular automata models that are not IBMs. They are treated here as one type, because IBMs are frequently based on cellular automata models.

Advantages of this type of model:

- Are able to account for individuality.
- Is able to account for adaptation within the spectrum of properties.
- Software is available; although the choice is more limited than by biogeochemical dynamic models.
- Spatial distribution can be covered.

The disadvantages are:

- If many properties are considered, the models get very complex.
- Can be used to cover the individuality of populations; but they cannot – at least not always – cover mass and energy transfer based on the conservation principle.
- Require many data to calibrate and validate the models.

As mentioned under the characteristics above, we know that the individuals have different properties and that may sometimes be crucial for the model results. In such cases, IBMs are absolutely needed.

2.3.9 Ecotoxicological models

Ecotoxicological models are in principle not representing a model type, as biogeochemical models or population dynamic models are applied widely in ecotoxicology. It is, however, preferable to treat ecotoxicological models as a separate model type, because they are characterized by:

1. Our knowledge to the parameters is limited and estimation methods are therefore needed to a much higher extent than for other model types. Fortunately, many estimation methods are available in ecotoxicology for estimation of process rates.
2. Due to the use of safety factors and the limited knowledge of the parameters, ecotoxicological models are often simple. Particularly, the so-called fugacity models illustrate this feature.
3. They should often include an effect component.

The advantages of this model type are:

- It is tailored to ecotoxicological problems.
- It is in most cases simple to use.
- It includes often an effect component or can easily be interpreted to quantify the effect.

The disadvantages are:

- The number of parameters needed to develop models for all toxic substances is very high and we know only at the most 1% of these parameters.

- It implies that we need estimation methods, which inevitably have a high uncertainty. The model results have, therefore, also a high uncertainty.
- Inclusion of an effect component requires knowledge to the effect, which is also limited.

The area of application is in this case obvious: to solve ecotoxicological research and management problems and perform environmental risk assessment for the application of chemicals.

2.3.10 Stochastic models

This model type is characterized by an element of randomness. The randomness could be the forcing functions, particularly the climatic forcing functions, or it could be the model parameters. The randomness is in both cases caused by a limitation in our knowledge. We can, for instance, not know the temperature on the 15th of May next year at a given location, but we know how the normal distribution of the temperature has been, for instance, since the last 100 years and can use the normal distribution to represent the temperature on this date. Similarly, many of the parameters in our models are dependent on random forcing functions or on factors that we hardly can include in our model without doing it too complex. A normal distribution of these parameters is known and by the use of a Monte Carlo simulation based on this knowledge, it is possible to consider the randomness. By running the model several times, it becomes possible to obtain the uncertainty of the model results. A stochastic model may be a biogeochemical/bioenergetic model, a spatial model, a SDM, an IBM or a population dynamic model. There are in principle no differences among these model types on how a model can be made a stochastic model.

This model type has the following advantages:

- This model type is able to consider the randomness of forcing functions or processes.
- The uncertainty of the model results are easily obtained by running the model many times.

This model type has the following disadvantages:

- It is necessary to know the distribution of the random model elements.
- The model has a high complexity and requires much computer time.

It is recommended to apply stochastic models, whenever the randomness of forcing functions or processes are significant.

2.3.11 Hybrid models

It is in principle possible to develop hybrid models by any combination of two or more of the previously listed 10 model types; but only very few hybrid models have been developed up to now. It is expected that many more will be developed in the future to combine some of the advantages and eliminate some of the disadvantages of the existing models. *Ecological Modelling* has published several hybrid models that combine, for instance, a biogeochemical/bioenergetic dynamic model with the other model type. The result by combining the biogeochemical dynamic model and ANN is a hybrid model that may have causality (opposite ANN) and is able to squeeze as much information out of the database as ANNs usually are able to do.

2.4 Applicability of the different model types

Which model types are recommended to solve which problems? What are the data requirements of the different model types? These are the obvious questions to answer already in the first step of the modelling development (see Fig. 1.3).

As mentioned in the introduction to this chapter, new model types were needed to solve specific model problems, of which most emerged in the early mid-1970s when ecological modelling started to be applied more extensively as a tool in environmental management. Biogeochemical/bioenergetic dynamic models and population dynamic models had some short-comings which the ecological modellers have tried to solve for the last 30 years mainly by the development of new model types. Today, the short-comings have at least partially been eliminated. Further improvements will be possible by the application of hybrid models or even completely new model types; but solutions to the problems formulated in Section 2.2 are available today. Consequently, it is possible to indicate which model type is the best choice in a given model situation, which is defined by the available data and the combination of problem and system. It is possible to indicate for the 11 model types in hand which type that it is recommended to use when we know the following two points that are determining the choice of model type:

1. The available data sets.
2. The combination of problem and system.

The core question: "Which model type should be applied in which context?" is answered in Tables 2.3 and 2.4, covering respectively different data sets and different problem/system combinations.

Today, we have a wide spectrum of model types available to solve a wide spectrum of relevant ecological problems, including a description of shifts in species compositions, ecotoxicological effects and spatial distributions and including the use of heterogeneous data sets and of uncertain data sets. The wide spectrum of models that we have available today is richly represented in the ecological modelling literature. A more comprehensive review of all the types with detailed illustration and ready-to-go software can be found in [2].

All the types are widely used today and our knowledge about the practical use of all the types makes it possible to answer the 10 questions listed in the beginning of the chapter. It does, of

Table 2.3: Model types applicable to different data sets.

Description of data set	The following model types can be applied
High quality, homogenous	Biogeochemical/bioenergetic dynamic models and population dynamic models
High quality, heterogeneous	ANN
Low quality, homogenous	Steady state models
Typical or average values	Steady state models
Extreme values only known	Steady state models
Uncertain data	Fuzzy models
No data, only rules	Fuzzy models
Qualitative data only	Fuzzy models

Table 2.4: Model types applicable to different problems/systems.

Problem/system	Model types to be recommended
Exchange of matter or energy	Biogeochemical/bioenergetic dynamic models, ANN can be applied when causality is unimportant
Development of populations	Population dynamic models, ANN can be applied when causality is unimportant
Toxic compounds: distribution and effect	Ecotoxicological models
Individuality important for the model results	IBM
Adaptation decisive process	SDMs
Structural changes may occur	SDMs
Forcing functions and/or processes stochastic	Stochastic models
Spatial distribution important	Spatial models

course, not exclude the possibilities to develop new types of models or make hybrid to the existing types and in the future, to come up with even better solutions for the increasing complexity of the demanding modelling tasks.

Illustration 2.1

Give an example of a case study that is best solved by use of the nine model types listed in Table 2.2. Describe the case study by the problem, the ecosystem and the data needed. Present the answer by use of a table.

Solution
The answers to the raised questions are answered in Table 2.5.

Table 2.5: Case studies illustrating the application of the various model types.

Model type	Description of case study
Dynamic biogeochemical models	Eutrophication of a lake, a relatively good database is available
Steady state biogeochemical models	Eutrophication of a lake, only bi-annual average values are available
Population dynamic models	Management of deer in a national park, a relatively good database is available
Spatial models	Distribution of nutrients in a landscape, a relatively good database is available
SDMs	Oxygen deficiency in a stream, significant changes of control functions are expected, knowledge about shifts in species composition is available

Table 2.5: Case studies illustrating the application of the various model types. *(Continued)*

Model type	Description of case study
SDMs	Oxygen deficiency in a stream, significant changes of control functions are expected, knowledge about shifts in species composition is available
IBMs and cellular automata	Growth of trees in a forest. The trees have very different conditions (sun, exposure to wind, soil etc.), a good database with different growth pattern under different conditions is available
ANN and use of artificial intelligence	The presence of different fish species in a wide spectrum of different streams. A huge but heterogeneous database is available
Fuzzy models	Presence or absence of five species of song birds in 20 different wetlands
Ecotoxicological models	The fate of an insecticide used in agriculture. An agricultural area, a wetland and a stream are considered

References

[1] Jørgensen, SE., *Integration of Ecosystem Theories: A Pattern*. 3rd edn, Kluwer Acad. Publ.: Dordrecht, Boston, London, 2002.

[2] Jørgensen, S.E., Chon, T. & Recknagel, F., *Handbook of Ecological Modelling and Informatics*, WIT Press: Southampton, 2009.

Exercises/problems – Chapter 2

1. Which type of model would you select for the following problems?
   (a) Protection of a lion population in a national park.
   (b) Optimization of fishery in a marine ecosystem, for instance, the North Sea.
   (c) Construction of a wetland for removal of nitrate mainly by denitrification.
   (d) Adaptation (change in the rate of evapotranspiration) of plants to a dry climate.
   (e) Interpretation of a database with 12 station in drainage areas of rivers and many observations as function of time for the stations of (i) water quality, (ii) fish diversity, (iii) dominant fish species and (iv) the use and composition of the land adjacent to the rivers.
2. In Illustration 1.1 two different process descriptions were presented: a first order reaction and a Michaelis–Menten expression. In which model types would these two descriptions under all circumstances be used? In which model types could the two descriptions be used? In which model types would the two descriptions under no circumstances be used?
3. A shallow lake with a surface of 100 ha and a depth of 2 m has an initial phosphorus concentration of 0.1 mg/l. The loading is 100 kg/year. The water retention time is 4 months. No input of phosphorus from the sediment is considered.
   (a) What would be the concentration of phosphorus in the lake at steady state, if it is considered that no settling of phytoplankton or suspended matter takes place?
   (b) What would be the concentration of phosphorus in the lake at steady state, if it is considered that phosphorus is settled by a rate of 1 m/24 h?
   (c) What would be the phytoplankton concentration in the two cases if it is considered that phytoplankton contains 1% phosphorus?

The differential equation needed to answer question (a) and (b) should be indicated and the steady state solution should be found.

CHAPTER 3

Biogeochemical models

3.1 Biogeochemical and bioenergetic models have wide applications

In the 1970s, almost two-thirds of all developed models were biogeochemical models, while one-third of all models today still belong to this model type. A biogeochemical model consists of the five model components presented in Chapter 1. A conceptual diagram shows how the selected state variables are connected and the forcing functions describe the exchange between the ecosystem and its environment. The differential equations set up a "book-keeping" for each state variable:

$$\text{Accumulation} = \text{inputs} - \text{outputs} \tag{3.1}$$

The inputs and outputs are processes that transfer mass between two state variables, or are forcing functions giving inputs and outputs between a state variable and the environment. The differential equations are straightforward to write down as shown in eqn (3.1), while the equations for the processes usually must be based on a good knowledge to the causality of the physical, chemical and ecological processes in ecosystems. The processes are described by algebraic equations, for instance, by first order processes where the process rate is proportional to the donor component or by a Michaelis–Menten expression to mention the may be two most applied equations in biogeochemical models. The process equations contain coefficients that have to be determined. They are denoted parameters and can be determined by:

1. use of literature values;
2. calibrating the model towards observations;
3. use of ecological knowledge (equations, principles), for instance, the allometric principles or mass balances (for illustrations of this method, see Chapter 5);
4. use of experience from previous case studies;
5. use of knowledge about the state variables at steady state conditions, which yield algebraic equations, that can be used to find the parameters;
6. use of intensive/very frequent measurement, it is presented in Section 3.4.

As the determination of the right set of parameters is always a weak point in the application of models [1], it is recommended to use, if possible, all six parameter determination methods simultaneously.

The forcing functions on the other hand require the determination of the rates of exchange between a considered ecosystem and its environment. The forcing functions are indicated as tables, graphs or equations.

Bioenergetic models are in principle very similar to biogeochemical models. Instead of transfer of mass they are focusing on energy "book-keeping". Organic matter contains chemical energy and chemical energy is stored in the form of organic matter. It implies that the differences between biogeochemical and bioenergetic models are minor.

These types of models have found a wide application, because pollution is a matter of input of polluting components (or energy) from the environment to ecosystems. The environmental management is concerned with the harmful effect the pollution is able to cause in the ecosystem, which can be revealed by a biogeochemical or bioenergetic model. The models applied in environmental management are, therefore, often biogeochemical or bioenergetic models.

Most ecological processes that are important for understanding of the ecology of ecosystems involve transfer of mass and/or energy. It also explains that ecological models applied in ecological research are frequently biogeochemical or bioenergetic models. The biogeochemical models attempt to set up relationships between external and internal variables – between forcing functions and state variables. The relationships are several mathematical equations.

Biogeochemical models are very powerful tools when good observations and detailed knowledge about the processes for a specific ecosystem are available. With good observations in hand and a detailed knowledge about the parameters, it will be possible to make a good calibration of the model. If the model can be validated with good result, the developed model is highly applicable in environmental management and research. If on the other side, the observations are scarce and the knowledge particularly about the parameters is limited, the model results will most probably have a high standard deviation, which inevitably limits the applicability of the model. In this case, it is still recommendable to develop a model, but the model results should be used carefully and sometimes only qualitatively or it could be considered to apply a steady state model instead of a dynamic model. In other words, a good data set and good knowledge about the crucial processes and parameters are the prerequisites for a dynamic biogeochemical model that can be used to make reliable prognoses. This discussion also shows that the validation of the model by an independent data set is extremely important, because the validation gives information about the standard deviation between the model results and the observations and thereby about the probable standard deviation of the prognoses. The development of biogeochemical models requires that the procedure presented in Chapter 1 under all circumstances is followed very carefully.

Biogeochemical models presume that the model focuses on a specific ecosystem in relation to a specific problem (see Chapters 1 and 2). Due to high sensitivity of at least some of the parameters it is obvious that the model presupposes that the parameters express the properties of the components in a specific ecosystem. The same biogeochemical model may be applied on several case studies with or without modification; but observations from every modelled ecosystem should be available to allow the modeller to determine the most sensitive parameters for the different ecosystems. This rule may have a few exceptions, when, for instance, a biogeochemical model is used to describe how a number of the same type of ecosystems will react to certain impacts. If the model is simple and the most sensitive parameters can be given almost the same values for all the considered ecosystems, it will be possible to apply a biogeochemical model for a number of ecosystems of the same type. The standard deviation is of course higher for such an application than for an application of the model for a specific ecosystem supported by a good data set.

The more complex a biogeochemical model is the higher is, of course, the probability that the model includes; a process which is very specific for the modelled ecosystem. The generality of biogeochemical models decreases, therefore, with increasing model complexity, expressed as number of state variables and/or by the number of parameters.

To summarize: if

1. the available data set is of good quality and gives sufficient information for the foreseen biogeochemical model,
2. a specific problem and a specific ecosystem are considered,
3. a model with high causality is desirable,

a biochemical model is a very good solution to the problem.

Thousands of biogeochemical models have been developed. Information of more than 400 selected examples can be found in Jørgensen *et al.* [2]. This model type has found a wide application for a number of different ecosystems and different environmental problems. Software is available for the development of this type of models, for instance, STELLA and SIMULA. Eutrophication models with increasing complexity, for shallow and deep lakes and for estimation of the nutrient input to a lake from the drainage area, can be found on UNEP's home page. The models that can be downloaded from this home page include a structurally dynamic model (see Chapter 7). Wetland models are or will also be available on the same home page.

Five examples are presented in the following sections to illustrate how biogeochemical models are developed and which considerations are necessary to make to develop this type of models properly. One steady state model is shown – the so-called chemo-state model. It will illustrate the advantages and disadvantages by a steady state model compared with a dynamic biogeochemical model. Two eutrophication models are presented. The first one is a simple model with only four state variables. The model can be found ready to use in the software Pamolare, which is downloadable from UNEP's home page. The model illustrates very clearly what can be obtained by the use of a simple model, but it also shows the considerations that the modeller inevitably will need to make when he wants to improve the model by adding more state variables.

The second eutrophication model demonstrates what can be obtained by a model with a medium to high complexity and also which comprehensive considerations are necessary to make for the choice of process equations and parameters. The eutrophication model has in this case been used to set up prognoses, which have even been validated. The model illustrates, therefore, the accuracy that can be expected of prognoses made on the basis of biogeochemical models that have been supported by a good database.

As an illustration for biogeochemical models, a model of a subsurface wetland has been selected. The model was developed as a result of a DANIDA project in Tanzania with the objectives to generate local know-how and expertise in the use of waste stabilization ponds and constructed wetlands for the treatment of wastewater. It was an integrated part of this project to develop a model that could be applied for the design of constructed subsurface wetlands. The model has been used in Tanzania for the design of subsurface wetlands and UNEP has developed a user-friendly version in C++ to be used for the construction of subsurface wetlands. The model has been applied in Tanzania, in Brazil, in Iraq and in Canada for the construction of wetlands or the use of natural subsurface wetlands for the treatment of wastewater. The model is presented below in details as a clear and illustrative example of biogeochemical models.

Finally, the development of river models is discussed. A simple river model considering the decomposition of organic matter and nitrification and the influence of these two processes on the oxygen balance is presented in details. It is briefly discussed how the model could be expanded to cover a wide spectrum of relevant processes.

3.2 A chemo-state model to illustrate a steady state biogeochemical model

A chemo-state is presumed to be a mixed flow reactor see Fig. 3.1, which implies that the concentrations of all the components are the same in the entire reactor. There are, in other words, no concentration differences or gradients in the reactor. Physical, chemical and biological processes take place in the reactor, but the reactor is sufficiently well mixed to maintain the same concentration throughout the entire reactor. A well-mixed aquatic ecosystem can be considered as a mixed flow reactor or a chemo state. The concentrations of the various components can be found as function of the time by use of the developed system of equations for this case. It is easy to find the steady state concentrations, which will give a good picture of the conditions in the aquatic ecosystem.

Let us consider one component in a chemo-state with the concentration, C. The following differential equation is valid:

$$dC/dt = (\text{input} - \text{output} - \text{decomposition} - \text{settling} - \text{evaporation})/V \quad (3.2)$$

where V is the volume (m^3). The processes can be expressed by algebraic equations, which change the differential equation to:

$$dC/dt = (I - Q \times C)/V - k \times C - sr \times C - A \times v \times C/V)\ mg/(m^3/24\ h) \quad (3.3)$$

where I is the input in mg/24 h, Q is the flow rate (m^3/24 h), k is the first order decomposition rate coefficient for the component, sr is the settling rate (1/24 h), A is the surface area (m^2) and v is the evaporation rate (m/24h) of the component. At steady state, the following equation is valid:

Steady state: dC/dt = 0, which implies that

$$I = (Q \times C + V \times k \times C + V \times sr \times C + A \times v \times C) \quad (3.4)$$

If I, Q, V, A, k, sr and v are known, C can be expressed as function of time and the steady state value can be found:

$$C = (I - Q \times C)/V - k \times C - sr \times C + A \times v \times C/V) \quad (3.5)$$

Illustration 3.1

A pesticide (Pz) is discharged to a lake, which can be considered a mixed flow reactor. 200 m^3 of wastewater with a Pz-concentration of 2 mg/l is discharged into the lake per 24 hours.

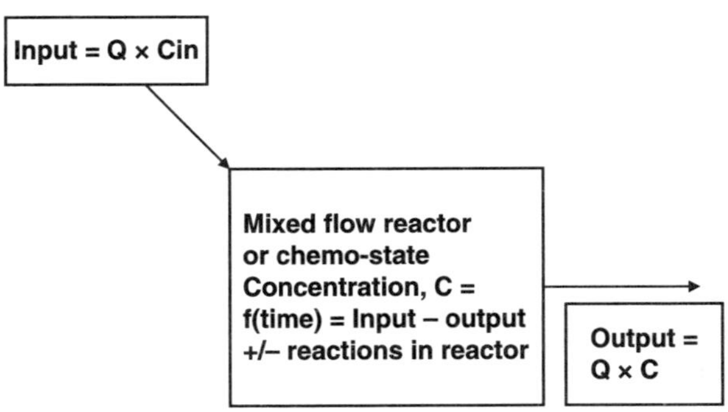

Figure 3.1: The principle for a mixed flow reactor or chemo-state is shown.

The natural inflowing stream has a rate of $800\,m^3/24$ h. The pesticide has a biological half-life time of 30 days. The settling and the evaporation are negligible. The volume of the lake is $100{,}000\,m^3$. What is the steady state concentration of Pz?

Solution

The specific decomposition rate k can be found from the first order differential equation for the decomposition:

$$dPz/dt = k \times Pz$$

which means since the half-life is 30 days:

$$\ln 2 = 0.693 = k \times 30 \text{ or } k = 0.0231 \text{ (1/24 h)}$$

The differential equation: dPz/dt = (input – output – decomposition)/V, which implies that:

$$dPz/dt = (200 \times 2 - Pz \times 1000)/100{,}000 - 0.0231 \times Pz = 0 \text{ (at steady state)}$$

or

$$0.004 - 0.01 \times Pz - 0.0231Pz = 0; \text{ or } \underline{Pz = 0.12\,mg/m^3}$$

3.3 An eutrophication model based on 2–4 state variables and statistic correlations to find additional state variables

From a thermodynamic view, a lake is an open system, which exchanges material (wastewater, evaporation and precipitation) and energy (evaporation and radiation) with the environment. However, in some great lakes the input of material per year is not able to change the concentration measurably. In such cases, the system can be considered as almost closed, which means that it exchanges energy, but not material with the environment. The flow of energy through the lake system leads to at least one cycle of material in the system [3]. As illustrated in Fig. 1.4, important elements can participate in cycles that control eutrophication.

The word eutrophy is generally taken to mean "nutrient rich". Nauman introduced in 1919 the concepts of oligotrophy and eutrophy, distinguishing between oligotrophic lakes containing little planktonic algae and eutrophic lakes containing much phytoplankton. The eutrophication of lakes all over the world has increased rapidly during the last decade due to increased urbanization and consequently increased discharge of nutrient per capita. The production of fertilizers has grown exponentially in this century and the concentration of phosphorus in many lakes reflects this.

The word eutrophication is used increasingly in the sense of artificial addition of nutrients, mainly nitrogen and phosphorus, to water. Eutrophication is generally considered to be undesirable, but this is not always true. The green color of eutrophied lakes makes swimming and boating less safe due to the increased turbidity, and from an aesthetic point of view the chlorophyll concentration should not exceed $100\,mg/m^3$. However, the most critical effect from an ecological point of view is the reduced oxygen content of the hypolimnion, caused by the decomposition of dead algae. Eutrophic lakes sometimes show a high oxygen concentration, during the summer time, at the surface, but a low concentration of oxygen in the hypolimnion, that is lethal to fish.

About 16–20 elements are necessary for the growth of freshwater plants, and Table 3.1 lists the relative quantities of the most important elements in plant tissue. The composition is on basis of wet weight. For plant tissue, water is 75–90% of the total wet weight. It means that except for oxygen and hydrogen, the composition on dry weight basis would be 8–10 times higher. For phytoplankton, the contents of carbon, nitrogen and phosphorus on dry weight basis are respectively approximately 40–60%, 6–8% and 0.75–1.0%. Phosphorus is considered to be the major cause of eutrophication in lakes, as it was formerly the growth-limiting factor for algae in the majority of lakes. Its use has increased tremendously during the last decade.

Nitrogen is limiting in a number of East–African lakes as a result of the nitrogen depletion of soils by intensive erosion in the past. However, today nitrogen may become limiting in lakes as a result of the tremendous increase in the phosphorus concentration caused by discharge of wastewater, which contains relatively more phosphorus than nitrogen. While algae use 5–10 times more nitrogen than phosphorus, wastewater generally contains only three times as much nitrogen as phosphorus. In lakes, a considerable amount of nitrogen is lost by denitrification (nitrate-> N_2).

The growth of phytoplankton is the key process of eutrophication and it is therefore important to understand the interacting processes that regulate growth. Primary production has been measured in great detail in a number of lakes. This process represents the synthesis of organic matter and the overall process can be summarized as follows:

$$\text{Light} + 6CO_2 + 6H_2O \rightarrow C_6H_{12}0_6 + 60_2$$

The composition of phytoplankton is not constant (note that Table 3.1 gives only an average concentration). The composition of phytoplankton or plants in general reflects to a certain extent the concentration of the water. If, for example, the phosphorus concentration is high, the phytoplankton will take up relatively more phosphorus – this is called luxury uptake.

As shown in Table 3.1, phytoplankton consists mainly of carbon, oxygen, hydrogen, nitrogen and phosphorus: without these elements no algal growth will take place. This leads to the concept of the limiting nutrient which has been developed by Liebig as the law of the minimum.

Table 3.1: Average freshwater plant composition on a wet weight basis.

Element	Plant content %
Oxygen	80.5
Hydrogen	9.7
Carbon	6.5
Silicon	1.3
Nitrogen	0.7
Calcium	0.4
Potassium	0.3
Phosphorus	0.08
Magnesium	0.07
Sodium	0.04
Iron	0.02
Manganese	0.0007
Zinc	0.0003
Copper	0.0001

This states that the yield of any organism is determined by the abundance of the substance that in relation to the needs of the organism is least abundant in the environment [4]. However, the concept has been considerably misused due to oversimplification. First of all, growth might be limited by more than one nutrient. The composition as mentioned above is not constant, but varies with the composition of the environment. Furthermore, growth is not at its maximum rate until the nutrients are used, and is then stopped, but the growth rate slows down when the nutrients become depleted.

It is discussed in more detail in the next section. It is also discussed how the interaction of several limiting nutrients simultaneously can be taken into account. Another side of the problem is the consideration of the nutrient sources. It is of importance to set up mass balances for the most essential nutrients. The best environmental strategy is not necessarily to limit the limiting nutrient to abate eutrophication, but to limit the nutrient which easily can be made the limiting nutrient. It is, for instance, often more easy to limit the phosphorus input to an aquatic ecosystem, because it is discharged mainly as wastewater, from where it can be removed in high efficiency by chemical precipitation, while nitrogen is present in agricultural drainage water, which is much more difficult to treat, of course.

Oligotrophic waters will usually have a ratio of N:P greater than or equal to 10, which means that phosphorus is less abundant than nitrogen for the needs of phytoplankton. If sewage is discharged into the lake the ratio will decrease, since the N:P ratio for municipal wastewater is 3:1, and consequently nitrogen will be less abundant than phosphorus relative to the needs of phytoplankton. In this situation, however, the best remedy for the excessive growth of algae is not the removal of nitrogen from the sewage, because the mass balance might then show that nitrogen-fixing algae will give an uncontrollable input of nitrogen into the lake. It is necessary to set up mass balances for each of the nutrients and these will often reveal that the input of nitrogen from nitrogen-fixing blue green algae, precipitation and tributaries is contributing too much to the mass balance for the removal of nitrogen from the sewage to have any effect. On the other hand, the mass balance may reveal that the phosphorus input (often more than 95%) comes mainly from sewage, which means that it is indeed better management to remove phosphorus from the sewage than nitrogen. Thus, as underlined above, it is not important which nutrient is most limiting, but which nutrient can most easily be made to limit the algal growth.

The development of a simple eutrophication model with let us say 2–4 state variables must obviously be based on the most important state variables. The immediate shot at the most important state variable for a eutrophication model is phytoplankton, but the experience has shown that the limiting nutrient in the water phase is the most important state variable. Phytoplankton is not the second most important state variable – at least not for shallow lakes. The sediment usually has high concentrations of nutrients, which are important to include to describe properly the dynamics of the exchange of nutrients between water and sediment than to consider the phytoplankton growth. These considerations imply that the following two differential equations seem to be the best approach for a two state variable eutrophication model with one limiting nutrient – in this case phosphorus:

$$dPS/dt = Q \times (P_{in} - PS)/V - s \times PS + rl \times P_{sed} \times al/d \tag{3.6}$$

$$dP_{sed}/dt = k \times s \times PS \times d/al - rl \times P_{sed} \tag{3.7}$$

where PS is the total phosphorus concentration in the water in g/m^3, Q is the inflow = outflow in $m^3/24$ h, P_{in} is the concentration of the inflow in g/m^3 as a mixture of wastewater, drainage water and tributaries, s is the settling in 1/24 h (or sometimes unit 1/y is used) equal to the settling in m/24 h divided with the depth (denoted d and expressed in m), rl is the release rate coefficient for the phosphorus in the sediment (P_{sed}), al is the depth of the active layer of the sediment. Notice that the settling is calculated based on the total phosphorus, while it is, of course, only the part of the phosphorus which is in form of phytoplankton that can actually settle. It means that if on average only the fraction f of the total phosphorus is phosphorus in phytoplankton, the settling rate should be f × the real settling rate of phytoplankton, which for most phytoplankton species is 0.05–0.5 m/24 h.

The ratio al/d gives the ratio in the volume of the active sediment layer and the water. al/d yields, therefore, the dilution of the phosphorus concentration when transferred from sediment to water. al is usually 5–15 cm and can be determined by the use of a sediment profile (see the discussion in Section 3.4), where a more complex sediment–water exchange submodel is proposed.

k is coefficient that considers that not all the phosphorus in the sediment can be released: k = ratio exchangeable P/total P. A part of the phosphorus is bound very firmly in the sediment and can never be released. The fraction 1–k will never be released but is mainly bound chemically as calcium, iron or other compounds. It is very essential to have a good estimation of k, because k is very decisive for the rate at which a lake will recover. Consequently, the prognoses become wrong if the approximately right k-value is not included in the model. The value k = 1 was applied for the very first developed eutrophication models and they yielded, of course, a very pessimistic prognosis for the recovery process, as the total amount of nutrients in the sediment could be released to the water. As it will be discussed in more detail in the next section, k is in the order of 0.6–0.9, which is able to accelerate the recovery process significantly.

Similar equations could be used for nitrogen. Translations of PS or NS, depending on which nutrient is limiting to phytoplankton, to the phytoplankton concentration as chlorophyll a (chl. a), primary production, zooplankton, fish production and transparency are possible by regression equations.

To illustrate the regression equations, a widely applied relationship between nutrient concentration and the average summer phytoplankton concentration will be used. Dillon and Rigler [5] developed a relationship for estimating the average summer chl. a concentration with the N:P ratio of the water greater than 12:

$$\log_{10}(\text{chl. a}) = 1.45\log_{10}[(\text{P}) \cdot 1000] - 1.14 \tag{3.8}$$

For the case where the N:P ratio is less than 4, the following equation based on eight case studies was evolved:

$$\log_{10}(\text{chl. a}) = 1.4\log_{10}[(\text{N}) \cdot 1000] - 1.9 \tag{3.9}$$

N and P are expressed as mgl^{-1} while chl. a is found in μgl^{-1}. If the N:P ratio is between 4 and 12, the use of the smallest value of chl. a found on the basis of the two equations is recommended. For the relationship between chl. a and the transparency, the graph shown in Fig. 3.2 is often applied. The graph presumes that the transparency is mainly dependent on the phytoplankton concentration. In case if either inorganic (clay) or organic matter (detritus) reduces significantly the transparency shown in the graph in Fig. 3.2 cannot be applied.

A model with the state variables PS, NS, P_{sed} and N_{sed} and the above mentioned regression equations is available as a simple one-layer model in the UNEP-software Pamolare. It is relatively fast to use this simple model. It is, therefore, often recommendable to use the model as a first modelling approach to gain first insights in the most crucial processes that are determining the eutrophication, before a more complex model is developed.

The model has been successfully applied on Lake Washington close to Seattle. Further details see [1] and Pamolare Software, developed by UNEP – IETC. The result of the model application is shown in Fig. 3.3. Diversion of the wastewater took place in 1967 and the phosphorus concentration was 25% of the previous value, after 5 years. The model was able to predict approximately the development of the observed phosphorus concentration shown in Fig. 3.3.

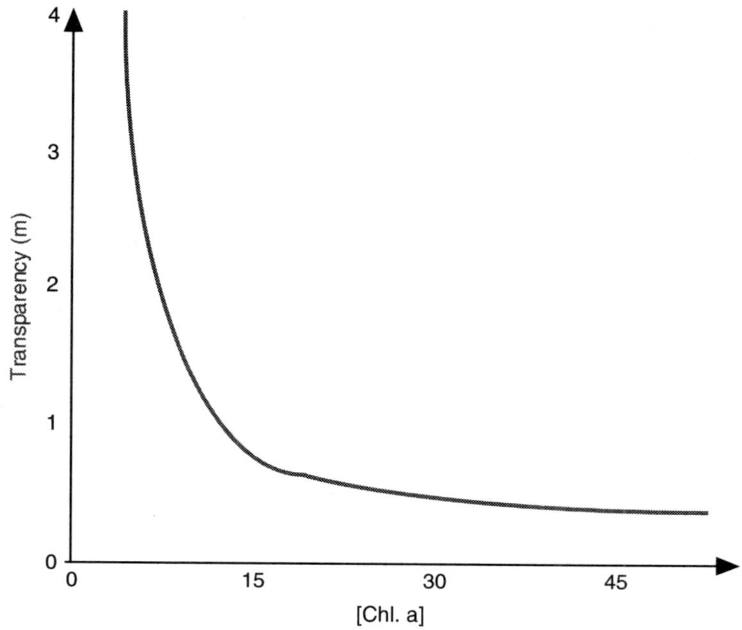

Figure 3.2: The relationship between chl. a (μg/l) and the transparency in m is shown.

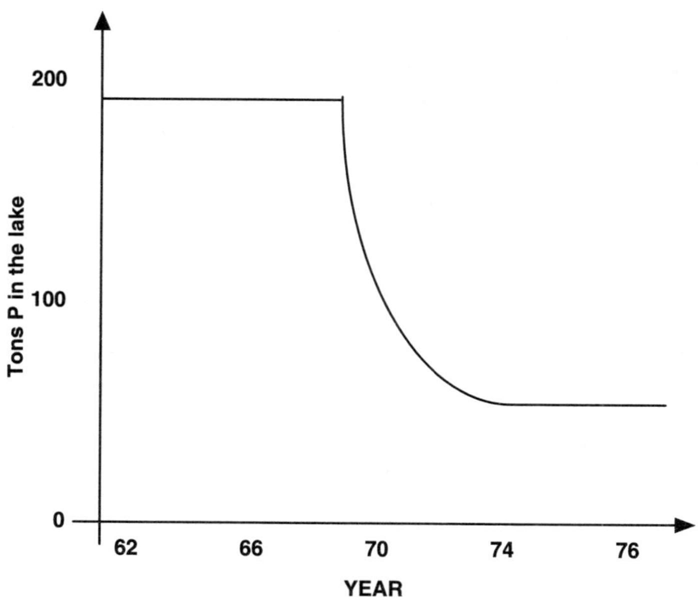

Figure 3.3: Diversion of the wastewater to Lake Washington took place in 1967. Five years after the diversion the lake was recovered and the amount of phosphorus in the lake reduced from almost 200 tons to about 50 tons.

3.4 A eutrophication model of medium to high complexity

There is a wide spectrum of eutrophication models available today:

1. From one to more than 70 state variables.
2. From one to seven layers, with and without thermocline (see Fig. 3.4 that shows recovery of a lake by treatment of hypolimnion water. The application of this ecotechnological method would inevitably require a model with at least two layers and thermocline).
3. From one to several hundred boxes.
4. Some of the models have been calibrated, validated and even the prognoses have been validated and some have even not been calibrated and validated, which, of course, is not acceptable for models that should be used in research or environmental management.
5. Some models use constant stoichiometric ratios of the nutrients, while other models are based on variable ratios of the nutrients. The difference is presented in more detail below.

As discussed in Chapter 1, it is not necessarily the most complex models that are the best models, because the right model complexity is a matter of a balance between the complexity of the problem, the ecosystem (the aquatic ecosystem) and the knowledge we have about the system and the problem including the quantity and the quality of data. The details of the model presented below, demonstrate clearly the considerations that are made when the model complexity is selected and changed (mainly increased). The model complexity should not be made blindly, for instance, just because a model of a certain complexity is available. Every change in complexity and particularly addition of complexity should be well considered.

The conceptual diagram shown in Fig. 1.4 is used for the nitrogen cycle of the model with a medium to high complexity presented below. The corresponding phosphorus cycle is shown in Fig. 3.5. The model is called the Glumsø Model, because it was applied for the first time on Glumsø Lake (Denmark).

It is clear from the conceptual diagrams that unlike the simple model presented in Section 3.3, phytoplankton and zooplankton are included as state variables. It implies that it has to be considered how the important processes phytoplankton growth and grazing of phytoplankton by zooplankton have to be formulated mathematically. Phytoplankton growth can be formulated as mentioned above by constant or by variable nutrient ratios. In the first case, the following mathematical formulations are usually applied:

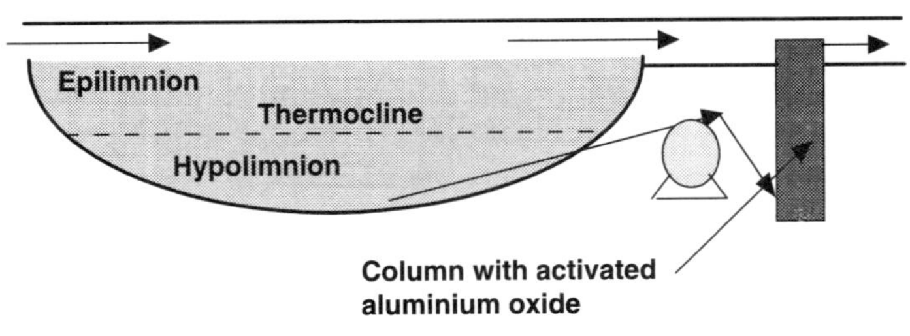

Figure 3.4: The thermocline separates the epilimnion and hypolimnion. The bottom water generally has a higher nutrient concentration than the epilimnion. It is therefore possible to recover the lake by discharge of the hypolimnion water, which, however, must be treated as shown in the figure, if the water can cause eutrophication downstream.

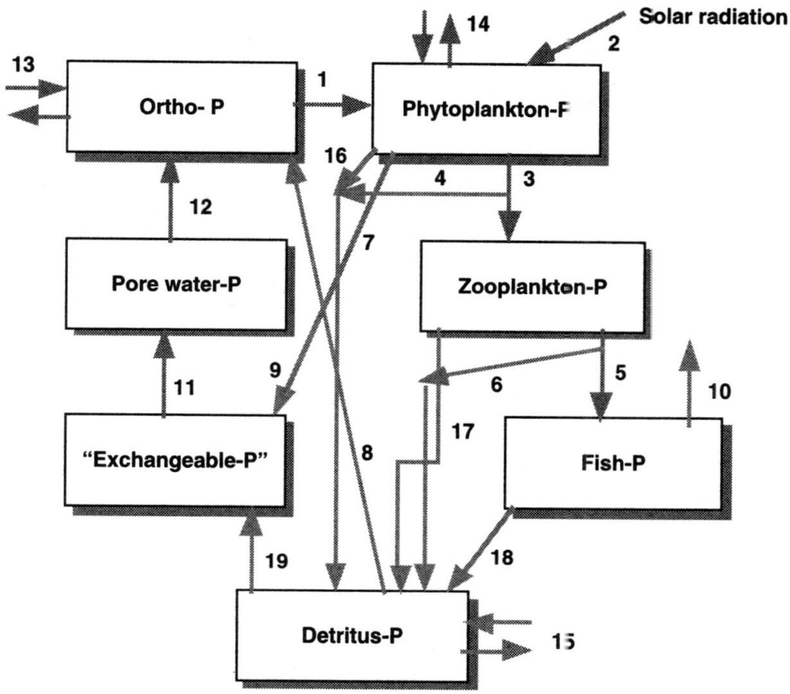

Figure 3.5: The phosphorus cycle. The processes are: 1 – uptake of phosphorus by algae; 2 – photosynthesis; 3 – grazing with loss of undigested matter; 4 and 5 – predation with loss of undigested material; 6, 7 and 9 – settling of phytoplankton; 8 – mineralization; 10 – fishery; 11 – mineralization of phosphorous organic compounds in the sediment; 12 – diffusion of pore water P; 13, 14 and 15 – inputs/outputs; 16, 17 and 18 – mortalities; and 19 – settling of detritus. Reproduced from [1].

The so-called Michaelis-Menten equation is applied for the growth, gr = grmax NS/ (kn + NS), or gr = grmax PS/(kp + PS), dependent on which nutrient is limiting N or P. grmax, kn and kp are parameters. If both nutrients are limiting in different period of the year, the formulation is

$$gr = grmax \min(NS/(kn + NS), PS/(kp + PS)) \quad (3.10)$$

Product or average of several limiting factors have also been proposed and applied.

For the influence of the temperature there are two possible formulations:

$$K_t\text{^}(TEMP - 20) \text{ (the so-called Arrhenius equation)} \quad (3.11)$$

or

$$\exp(A \times (TEMP-OPT)/(TEMP-MAXTEMP)) \quad (3.12)$$

K_t is a parameter, which in most cases is between 1.04 and 1.06, in average 1.05. OPT is the optimum temperature for phytoplankton growth and MAXTEMP is the maximum temperature. OPT, MAXTEMP and A are all parameters that are different for different phytoplankton species. Equation (3.12) gives, in most cases, a more correct description of the temperature dependence, but it also contains more parameters that have to be determined. It is of course possible to include

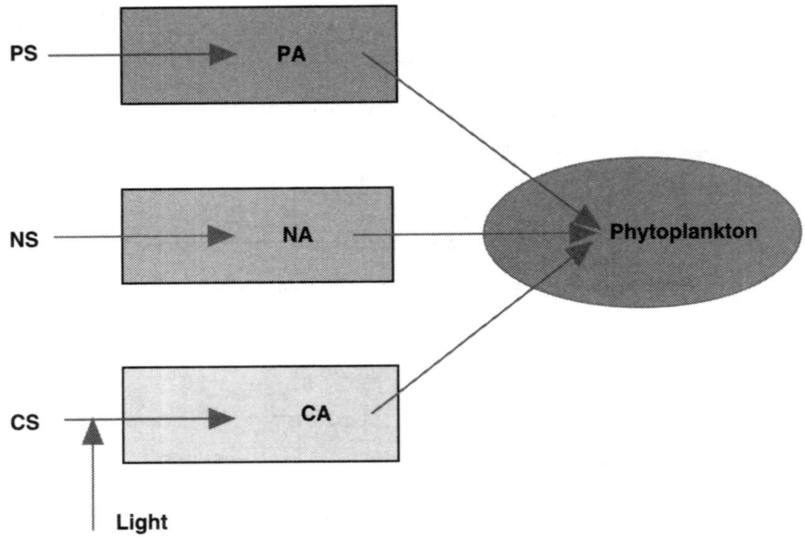

Figure 3.6: The two-step model of phytoplankton growth. The first step is uptake of nutrients PS, NS and CS, followed by a growth of phytoplankton depending on the nutrient concentrations in the phytoplankton cells, PA, NA and CA. The uptake of carbon is dependent on the light, while the uptake of phosphorus and nitrogen can take place even in darkness. It is a more physiologically correct description of phytoplankton growth than the eqns (3.8)–(3.10).

a limitation by silica and/or iron and other possible nutrients in a similar way as it is shown above for nitrogen and phosphorus.

The description of phytoplankton growth by the eqns (3.10)–(3.12) is a simplification, because the phytoplankton growth is in reality a two-step process, as illustrated in Fig. 3.6. The first step is uptake of nutrients and the second step is growth of phytoplankton (increase of the biomass). The more correct description can be formulated mathematically by the following equations:

$$\begin{aligned}&\text{Uptake rate P} = dPA/dt = PA \times maxupp \times (PS/(kp + PS))\\&\qquad\qquad \times ((PAMAX - PA)/(PAMAX - PAMIN))\\&\text{Parallel for uptake rate N}\end{aligned} \tag{3.13}$$

$$\begin{aligned}&\text{Uptake rate C} = dCA/dt = CA \times maxupc \times (CS/kc + CS)\\&\quad \times ((CAMAX - CA)/(CAMAX - CAMIN)) \times ((L/KL + L) - RESP)\\&\quad\quad \text{if } L < L1 \text{ L is used; If } L2 > L > L1 \text{ L1 is used;}\\&\quad\quad \text{if } L > L2 \; L1 + L2 - L \text{ is used for L}\end{aligned} \tag{3.14}$$

PA, NA and CA are state variables that cover the amount of phosphorus, nitrogen and carbon, in form of phytoplankton expressed as mg P, N or C per liter of water. Maxupp, maxupn, macups, kp, kn, kc, PAMAX, PAMIN, NAMAX, NAMIN, CAMAX, CAMIN, KL, L1 and L2 are all parameters. PAMAX, PAMIN, NAMAX, NAMIN, CAMAX and CAMIN are, however, known fairly well. They are the phytoplankton concentration times respectively 0.025, 0.005, 0.12, 0.05, 0.6 and 0.4 with good approximations. It is of course more difficult to calibrate the two-step growth equations than the eqns (3.10)–(3.12) due to the high number of parameters in the eqns (3.13) and (3.14), although the approximate knowledge to the six parameters, PAMAX, PAMIN,

NAMAX, NAMIN, CAMAX, CAMIN facilitates the calibration slightly. The uptake of phosphorus, nitrogen and carbon are according to the equations dependent on the concentrations of the nutrients in the water and of the concentrations of nutrients in the cells.

The closer the nutrient concentrations are to the minimum, the faster is the uptake. When a nutrient concentration has reached the maximum value the uptake stops. Notice that the carbon uptake unlike the uptake of phosphorus and nitrogen is also dependent on the light by a Michaelis–Menten expression that includes the light prohibition. At a light intensity L > L2, the growth is decreasing approximately linear according to the equation for uptake of carbon. At light intensity L1, the saturation level has been reached. A graph showing the growth as function of the light intensity is shown in Fig. 3.7. Finally RESP covers the respiration. Only carbon is, of course, involved in the respiration.

The growth process is quantified by the following equation:

$$\text{Growth} = \text{grmax} \times \text{phytoplankton} \times (\min((\text{PA} - \text{PAMIN})/(\text{PAMAX} - \text{PAMIN})), ((\text{NA} - \text{NAMIN})/(\text{NAMAX} - \text{NAMIN})), ((\text{CA} - \text{CAMIN})/(\text{CAMAX} - \text{CAMIN}))) \quad (3.15)$$

grmax is a parameter on line with corresponding parameter in eqn (3.10).

The phytoplankton growth model has four state variables: PA, NA, CA and phytoplankton. They are all assumed to apply the unit mg/l. As the minimum and maximum values are presumed to be a parameter times the phytoplankton concentration, they are also expressed in mg/l. It is in accordance with the discussion in Chapter 1 about the verification: check always the units.

The two-step description is, of course, more difficult to calibrate and validate and its use in general, which of course, raises the question: when should the two-steps description be applied instead of the easier applicable constant stoichiometric approach? As pointed out in the studies by Jørgensen [6–9], it is sometimes impossible to calibrate the one-step growth of phytoplankton,

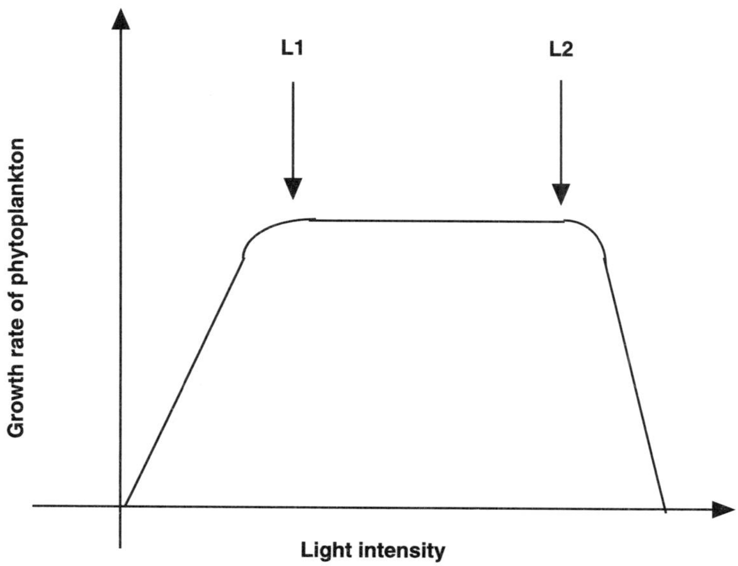

Figure 3.7: The phytoplankton growth is plotted versus the light intensity. At the light intensity L1, the saturation has been reached. At a light intensity greater than L2 photo prohibition takes place.

because the reaction rates at a high nutrient concentration are too fast. A general examination of many lake modelling case studies (see, for instance, the study by Di Toro and Conolly [10]) has revealed that the need for the two-steps description is increasing with the shallowness and the eutrophication of the aquatic ecosystem (see Fig. 3.8). It is definitely recommended for shallow very eutrophied aquatic ecosystems to apply the two-steps description, while it is hardly needed for deep mesotrophic or oligotrophic aquatic ecosystems.

As discussed in Section 3.3, the exchange of nutrients between sediment and water may be very important processes, because the amount of nutrients stored in the sediment is frequently several times bigger than the amount of nutrients in the water phase, particularly for shallow lakes. In the model presented here with a medium to high complexity, it is, therefore, natural to question the submodel for the nutrient exchange between sediment and water.

It has been found that the submodel shown in Fig. 3.9 for the exchange of nutrients between sediment and water, gives a proper description, which has been applied in the Glumsø Model that is presented in this section.

Sedimentation (S), corresponds to processes (9) and (19) in Fig. 3.5. A part of S is resuspended, particularly in shallow lakes. As discussed in Section 3.3, a part of the settled material is exchangeable and a part is non-exchangeable. The exchangeable part is decomposed (it is mainly organic material) and forms dissolved phosphorus in the pore water or interstitial water. A similar submodel can be applied for the nitrogen exchange between water and sediment. The dissolved nitrogen in the pore water is ammonium.

The parameter k (see eqn (3.7) and al (see eqns (3.6) and (3.7)) can be determined by use of a phosphorus and nitrogen sediment profile. Particularly k is different for phosphorus and nitrogen, but even the active layer may be different for the two nutrients. Figure 3.10 shows such a profile taken from (22). The active layer, al, is in this case 5 cm = LUL. k is B/(A + B) = 4.5/6.0 = 0.75. three-fourths or 75% of the top layer of the phosphorus in the sediment is able to return to the water phase, while one-fourth or 25% is bound very firmly chemically and cannot return to the water phase. As already mentioned, it is crucial for the prognoses of recovery of aquatic ecosystems

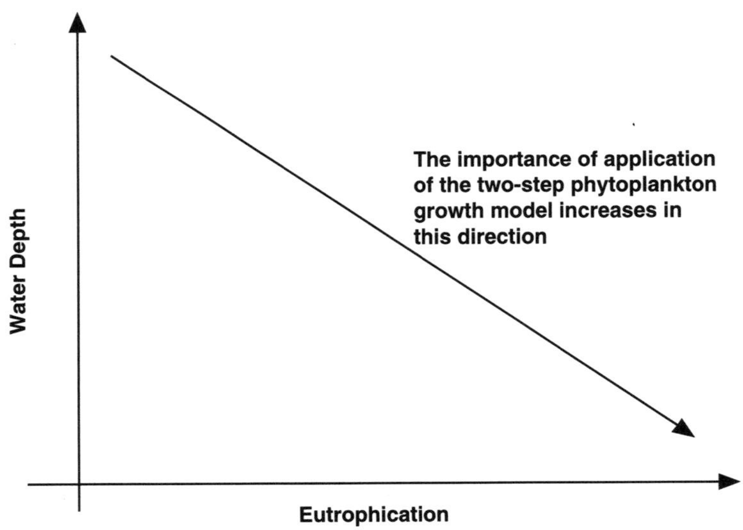

Figure 3.8: The need for the two-steps description of phytoplankton growth increases with the nutrient concentration and decreases with the depth.

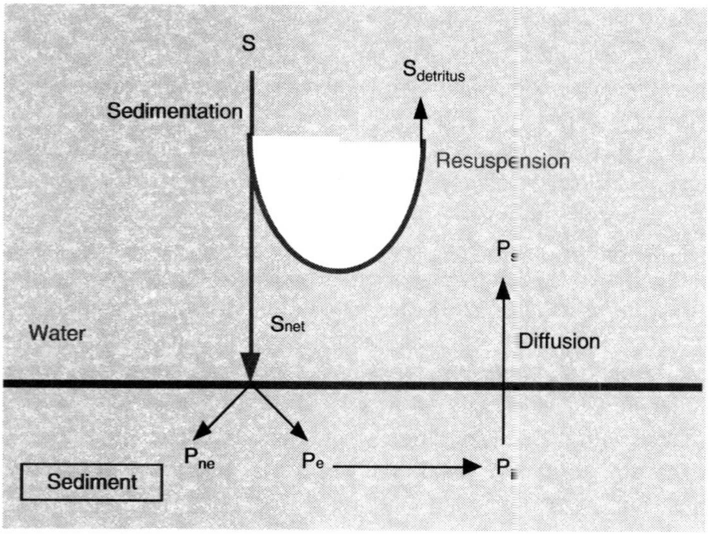

Figure 3.9: The sediment–water exchange submodel that is applied in the Glumsø eutrophication model. A part of the settled material is resuspended and a part of the net settled material is exchangeable and a part is non-exchangeable. The exchangeable part is decomposed (it is mainly organic) and forms dissolved phosphorus in the pore water (for dissolved nitrogen, ammonium is formed). The phosphorus (or for dissolved nitrogen, the ammonium) is by a diffusion process returned to the water phase. Reproduced from [1].

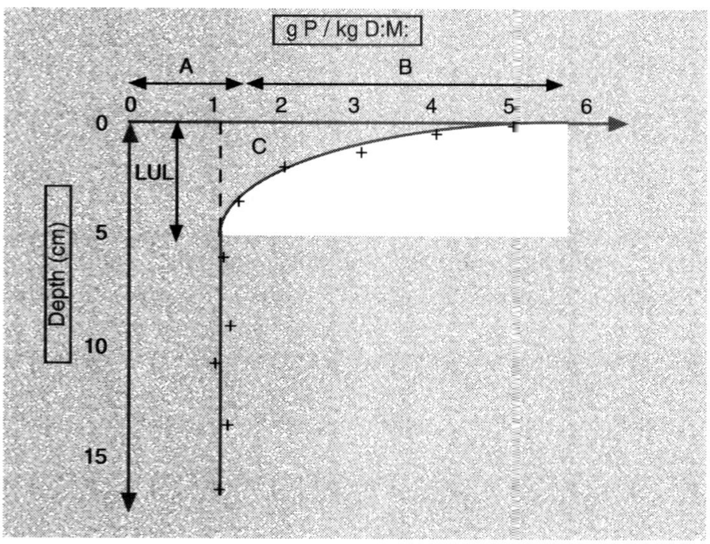

Figure 3.10: A phosphorus sediment profile for Lake Esrom in Denmark is shown. As it can be seen the active layer al – denoted LUL – is 5 cm and the ratio of exchangeable to total phosphorus in the sediment is k = 4.5/6.0 = 0.75. Particularly k is different for phosphorus and nitrogen, while the active layer, al, often is the same or approximately the same for phosphorus and nitrogen. Reproduced from [1].

after a significant eutrophication to account for the part of the nutrient that is not returned to the water phase. Usually k is higher for nitrogen than for phosphorus, because both ammonium and nitrate form soluble inorganic compounds, while there are several insoluble inorganic phosphorus compounds particularly with calcium and iron.

The grazing of phytoplankton by zooplankton can be expressed as a Michaelis–Menten equation with some modification. When the phytoplankton concentration is very low, no grazing is observed, because the zooplankton can only graze a certain volume per day. A threshold concentration is introduced to account for this situation.

On the other side, if the phytoplankton concentration is very high, zooplankton will grow too fast in accordance with the Michaelis–Menten expression, because it is no longer the amount of food that is limiting the zooplankton growth, but the carrying capacity, that is, the concentration of zooplankton that can be maintained in the aquatic ecosystem due to the predation by fish and the available shelter for protection of eggs and off-springs (nesting space). A regulation of the growth in accordance with the concentration of zooplankton relatively to the carrying capacity can be applied to account for this growth limitation. It is similar to what is used for a description of the so-called logistic growth (see Chapter 4).

According to the presented considerations, the following equation can be proposed for a proper description of grazing:

$$\begin{aligned}\text{Grazing} &= \text{Growth rate of zooplankton}\\ &= \text{zoo} \times \text{maxgraz} \times ((\text{Phyt} - \text{trc})/(\text{kph} + \text{Phyt})) \times (1 - \text{zoo}/\text{CC})\end{aligned} \tag{3.16}$$

where zoo is zooplankton, trc is the concentration below which there is no grazing and CC is the carrying capacity depending on the environmental conditions (available nesting space).

Similarly, the remaining processes were discussed: which equation would describe the nitrification? – a Michaelis–Menten equation was chosen based on the general experience. For the decomposition of detritus a first order reaction was selected, because it is in accordance with a fast regulation of a microbiological population to the available organic matter. It was, furthermore, attempted to introduce omnivorous, detrivorous and carnivorous zooplankton species, but it did not change the model results significantly. Therefore, the model based on the conceptual diagrams Figs 1.4 and 3.5 was selected to be used for Glumsø Lake to set up prognoses for the removal of phosphorus from the discharged wastewater. We will return to the validation of the prognoses later in this section. It is, however, necessary before the prognosis validation to discuss the use of very frequent measurements for a better determination of the most crucial parameters, the general validation and the general applicability of the model.

Figure 3.11 compares the results of very frequent measurement with measurements twice a month. It is clear from the graph that the latter frequency of measurements cannot give a proper picture of the dynamic in aquatic ecosystem, because phytoplankton is growing (changing) significantly on a day to day basis. Measurements every fortnight cannot reveal the dynamics of the eutrophication process. If the model is only supposed to give a monthly average, it may be sufficient to have observations only twice a month, but if we want to determine the peak value and reveal the dynamics at spring and summer bloom, it is clearly necessary to have more frequent measurements. For the development of eutrophication models it is therefore recommended to provide very frequent measurements when the spring bloom and/or summer bloom occurs. Also, it will be demonstrated below that such frequent measurements can be used to improve considerably the parameter estimation.

The frequent measurements can be applied to find a function of higher order, which is able to describe the phytoplankton = f(time). Statistical programs are available that can find a higher

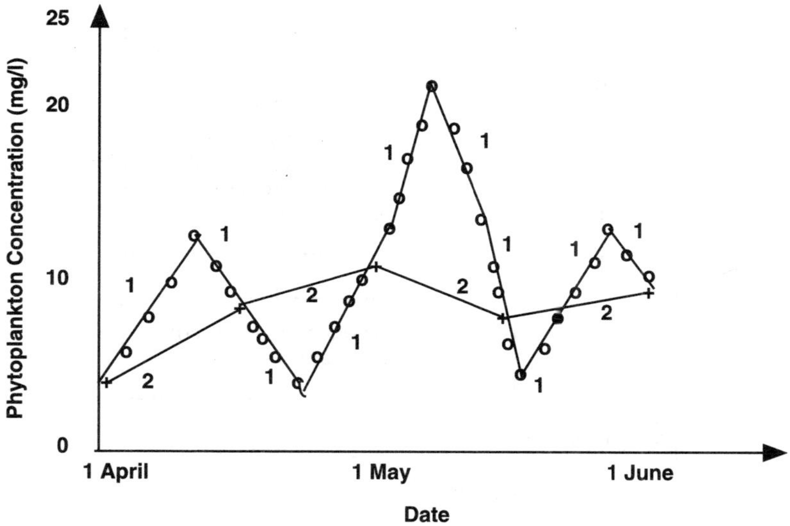

Figure 3.11: Measurement twice a month is compared with measurements four times a week for the period from 1st of April to mid-June for Glumsø Lake. The two measuring programs give clearly a different picture of the dynamic of phytoplankton. Reproduced from [1].

order algebraic expression that can follow the frequent measurement very closely. It means that we can also find dPhyt/dt the derivative as function of the time. If, for instance, phytoplankton = f(time) has been found to be:

$$\text{Phyt} = a + b \text{ time} + c \text{ time}^2 + d \text{ time}^3 + e \text{ time}^4 + f \text{ time}^5 \tag{3.17}$$

The phytoplankton growth as function of time will be:

$$\text{dPhyt/dt} = 0 + b + 2c \text{ time} + 3d \text{ time}^2 + 4e \text{ time}^3 + 5f \text{ time}^4 \tag{3.18}$$

It implies that eqn (3.18) for different values of time can be used to find unknown parameters. The model equation corresponding to (3.18) is:

$$\text{dPhyt/dt} = \text{maxgr} \times \text{Phyt} \times \min(\text{PS}/(\text{PS} + \text{kp}), \text{NS}/(\text{NS} + \text{kn})) - \text{sed} \times \text{Phyt} \ldots,$$

where … indicates that there may be additional expressions. Phyt, PS, NS and even dPhyt/dt are all known as function of time due to the very frequent measurements.

As unknown, we can use the four parameters maxgr, kp, kn and sed. If we have, for instance, 28 sets of observations of NS, PS, Phyt and therefore also of dPhyt/dt, we have 28 equations with only four unknown, which means that seven values for each of the four parameters can be obtained. It is therefore possible to calculate four average values of the four parameters and four standard deviations. These values are compared with the ranges published for these four parameters in the literature. If the values and their standard deviations are acceptable according to the comparison with the literature ranges, the average values can be used as initial guesses in the general calibration. If the values and/or the standard deviations are not acceptable, the model should be questioned. This application of the very frequent measurements is shown as a computer flow chart in Fig. 3.12.

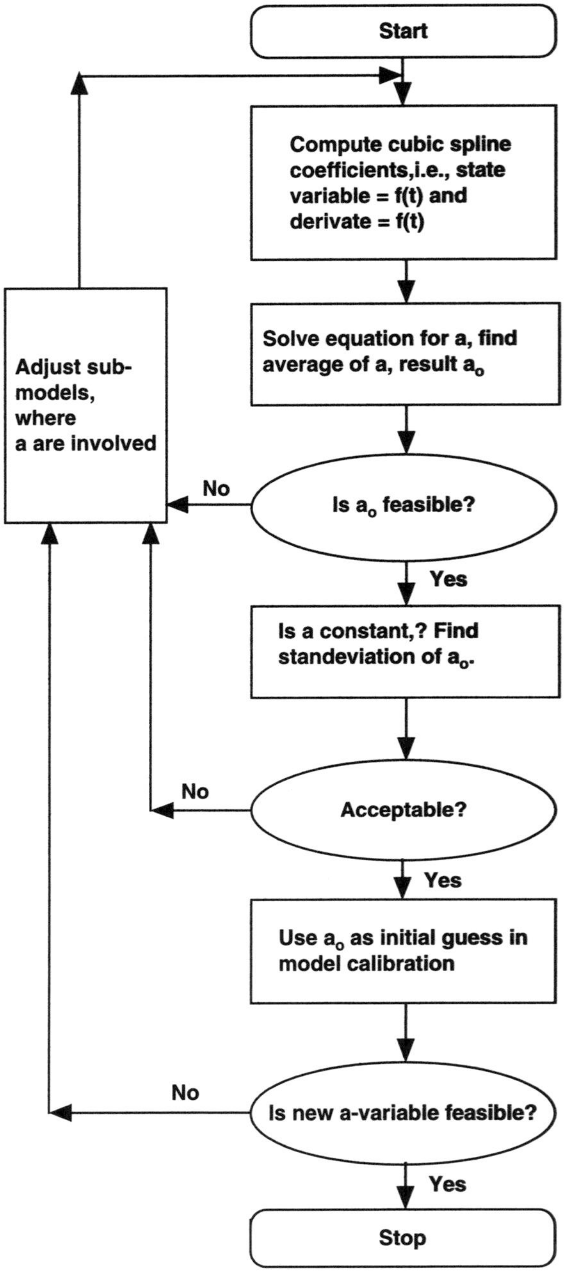

Figure 3.12: A computer flow chart that shows how frequent measurements can be applied to test the model and to obtain a better calibration. Reproduced from [1].

The use of frequent measurements for the development of the Glumsø Model gave as result, that eqn (3.12) should be applied to express the temperature dependence of the phytoplankton growth – not the simple Arrhenius expression shown in eqn (3.11). Furthermore, it was found that maxgr is 1.8 1/24 h ± 10% relatively, that is, between 1.62 and 1.98 1/24 h. It was very beneficial for the general calibration to know maxgr – a very important and sensitive parameter – with a standard deviation of ±10%. Due to a good performance of the calibration, a fully acceptable validation was achieved with the following results (relative percentage difference between modelled and observed values):

- The phytoplankton concentration (average of 15 comparisons) 10%.
- The peak phytoplankton concentration (1 determination) 15%.
- Production (gC/(m^2 24 h)) (average of 15 comparisons) 8%.
- Production peak value (gC/(m^2 24 h)) (1 determination) 5%
- All state variables (average of 225 comparisons) 16%.
- Zooplankton peak value (1 determination) 0%.
- The zooplankton concentration (average of 15 comparisons) 27%.
- Transparency (average of 15 comparisons) 15%.

The validation of the total assessment of the eutrophication of Glumsø Lake has a standard deviation of 16% divided by $\sqrt{225}$ or slightly more than 1%.

As the validation was acceptable, the model was applied to set up a prognosis for the production, phytoplankton concentration and transparency, if the phosphorus input to the lake was reduced by 90%, which was easy to achieve by a well-controlled chemical precipitation, for instance. Before the reduction of the phosphorus input, the lake was very eutrophied, which can be seen by the following typical observations:

- Total P g/m^3:1.1.
- Phytoplankton concentration peak value: (mg chl. a/m^3) 850.
- Production (gC/(m^2 year)) 1050.
- Minimum transparency at spring bloom (m) 0.18.

Fortunately, the water residence time of Glumsø Lake is only 6 months, which implied that it was possible to validate the prognosis properly during a few years. A comparison of the prognosis and the actual observations after 90% reduction of the phosphorus input is shown in Table 3.2. The indicated standard deviations in the table are for the prognosis based on the validation results shown above and for the measurement based on a general determination of the standard deviations for measurement on 10% relatively.

The prognosis validation is fully acceptable except for the daily production (gC/(m^2 24 h)) at spring bloom in the second year. In the beginning of the second year, the phytoplankton species shifted from *Scenedesmus* to various species of diatoms.

The shift implied that the well-determined parameters for phytoplankton were no longer valid, which may explain the discrepancy between the prognosis and the observations, particularly the second year after reduction of the phosphorus loading. The shift has, however, demonstrated clearly the need for a structurally dynamic modelling approach, as it will be discussed much further in Chapter 7 (see also [11]).

The ecological eutrophication model for Glumsø Lake has been applied on 23 other aquatic ecosystems with modifications. A model with a medium to high complexity is obviously less general than a simpler model, because it includes ecosystem specific processes. It is therefore not surprising that it may be necessary to modify and revise the model from case to case. An overview of the modifications is given in Table 3.3.

The Glumsø Model has been applied with modifications in totally 24 case studies. Most of the case studies have included calibration and validation (denoted level 5–6). Three case studies have

Table 3.2: Validation of the prognosis for Glumsø Lake.

Comparison of	Time	Prognosis		Measurements	
		Value	St. dev.	Value	St. dev.
Min. transparency	1st year	0.20 m	0.03	0.20 m	0.02
Min. transparency	2nd year	0.30 m	0.05	0.25 m	0.025
Min. transparency	3rd year	0.45 m	0.07	0.50 m	0.05
Max. production	2nd year	6.0 gC/24 hm^2	0.3	11.0 gC/24 hm^2	1.1
Max. production	3rd year	5.0 gC/24 hm^2	0.3	6.2 gC/24 hm^2	0.6
Max. chl. a	1st year	750 mg/m^3	112	800 mg/m^3	80
Max. chl. a	2nd year	520 mg/m^3	78	550 mg/m^3	55
Max. chl. a	3rd year	320 mg/m^3	48	380 mg/m^3	38
Annual production	2nd year	720 gC/y m^2	$15^{+)}$	750 gC/y m^3	19^{+}
Annual production	3rd year	650 gC/y m^3	$13^{+)}$	670 gC/y m^3	17^{+}

$^{+}$as standard deviation is used 8% for the prognosis divided by √15 and 10% divided by √15 for the measurements, because the determination of the annual production is based on 15 measurements and 15 prognosis values.

Table 3.3: Revision of the Glumsø eutrophication model by applications in other case studies.

Case studies	Revisions made
Lake Victoria, Lake Kyoga, and Lake Edward, Africa	Thermocline, other food chain, boxes
Fure Lake, Denmark	Nitrogen fixation, SDM, restoration
Ringkøbing Fjord, Denmark	Nitrogen fixation, boxes
Esrom Lake, Denmark	Silica included as nutrient
Lyngby Lake, Denmark	Recalibration only
Four lakes in Zealand, Denmark	Resuspension
Lake Gyrstinge, Denmark	Level fluctuations, sediment exposed to air
Roskilde Fjord, Denmark	Complex hydrodynamics, SDM
The Lagoon of Venice, Italy	Ulva and Zostera included, SDM
Stadsgraven and the internal lakes, Copenhagen	Interconnected basins
Søbygaard Lake and Veng Lake, Denmark	SDM
Lake Annone, Italy	SDM and two connected basins
Lake Mogan, Turkey	SDM, only P- cycle
Biomanipulation, general	SDM
Bergunda Lake, Sweden	Mixture of data from different years
Broia Reservoir, Brazil	Two different types of zooplankton

given occasion to validate prognoses (denoted level 7), while two case studies had to stop before the calibration due to lack of data of sufficient quality. Table 3.4 gives an overview of the number of case studies at different development levels.

The following indication of the levels has been applied:

- Level 1: Conceptual diagram developed.
- Level 2: Verification carried out.

Table 3.4: Levels of development for the 24 case studies where the Glumsø Model has been applied.

Level	Number of case studies
1: Only conceptual diagram	0
2: Verification made	1
3: Verification of sensitivity analysis made	1
4: Calibration made by either use of very frequent measurements or by use of annual measurements	3
5: Two calibrations are made	5
6: Calibration and validation	11
7: Validation of published prognoses	3

- Level 3: Sensitivity analysis carried out.
- Level 4: One calibration by either very frequent or annual measurements carried out.
- Level 5: Two calibrations are made – in most cases by very frequent and annual measurements.
- Level 6: Two calibrations and validation carried out.
- Level 7: Validation of the prognoses was possible.

Illustration 3.2

How would you change the state variable zooplankton and the grazing equation in case that:

(a) Zooplankton was omnivorous feeding on phytoplankton and detritus?
(b) We had to include two types of zooplankton, one using detritus as food source and one using phytoplankton as food source.
(c) Both a herbivorous and a carnivorous zooplankton should be included.

Solution
Growth rate of zooplankton = zoo × maxgraz × ((Phyt – trc)/(kph + Phyt)) × (1 – zoo/CC) would be changed to:

(a) Growth rate of zooplankton = zoo × maxgraz × ((Phyt + Detr – trc)/(kpd + Detr + Phyt)) × (1 – zoo/CC)
(b) Growth rate of zooplankton 1 = zoo1 × maxgraz1 × ((Phyt – trc)/(kph + Phyt)) × (1–zoo1/CC) and
Growth rate of zooplankton 2 = zoo2 × maxgraz2 × ((Detr – trc)/(kdr + Detr)) × (1–zoo2/CC)
(c) Growth rate of zooplankton 1 = zoo1 × maxgraz1 × ((Phyt–trc1)/(kph + Phyt)) × (1–zoo1/CC) and
Growth rate of zooplankton 2 = zoo2 × maxgraz2 × ((zoo1–trc2)/(kzo + zoo1)) × (1–zoo2/CC)

In both the equations, it could be considered to let the carrying capacity be determined by the sum of zoo1 and zoo2, which means that in both the equations (1 – (zoo1 + zoo2)/CC) is applied to account for the adjustment to the carrying capacity.

The predation by fish will take place on both zoo1 and zoo2, which means that the predation on zooplankton by fish would be expressed by the following equation:

$$\begin{aligned}\text{Growth of fish} &= \text{predation on zooplankton by fish}\\ &= \text{fish} \times \text{maxpred} \times (\text{zooa} + \text{zoo2} - \text{trcpred})/(\text{kzo1–2} + \text{zoo1} + \text{zoo2})\\ &\quad \times (1 - \text{fish/CCfish})\end{aligned}$$

kpd, kdr, kzo and kzo1–2 are all new half-saturation constants; maxgraz1, maxgraz2 and maxpred are all max growth rates; and trc1, trc2 and trcpred are all threshold concentrations.

All these modifications have been tested in the Glumsø Model with no or little changes of the results.

3.5 Model of subsurface wetland

The model presented in this section illustrates very clearly the basic ideas behind biogeochemical models. It was developed as a result of a DANIDA project promoting the cooperation between Copenhagen and Dar es Salaam University in Tanzania. Later, United Nations Environmental Program – International Environmental Technology Center (UNEP–IETC) developed a software based on the model, which could be used by developing countries to design subsurface wetlands. The software was called SubWet. Fleming College, the Center for alternative wastewater treatment, wanted to use the software for design of wetlands also in cold climate and supported the further development of the software to be applied both for warm and cold climate. This version of the software is denoted SubWet 2.0 see the home page of UNEP–EITC and [12].

The scope of the model is to design and manage a subsurface wetland, based on defined removal efficiencies of organic matter (expressed in terms of BOD_5), nitrate, ammonium, organic nitrogen and phosphorus. Thus, it is necessary to know:

1. the water flow;
2. the concentrations of the above mentioned constituents in the water; and
3. the required removal efficiencies for these constituents (i.e., their concentrations in the treated water).

The subsurface wetland being modelled consists of a constructed or a natural wetland area. The constructed wetland has gravel as soil, whereby it is possible to ensure a good water flow through the wetland. The core design parameter is the area and the volume of the wetland (denoted as A and V).

The conceptual diagram of the model is presented in Figs 3.13–3.15. Figure 3.13 illustrates the organic matter submodel, while Fig. 3.14 illustrates the nitrogen submodel, with three different nitrogen compounds (organic nitrogen, ammonium and nitrate). Figure 3.15 illustrates the phosphorus submodel. The state variables of the model are: BOD_5, nitrate (NIT), ammonium (AMM), total phosphorus (TP) and organic nitrogen (ORN) in 5 successive boxes, denoted as A, B, C, D and E. Totally the model has 25 state variables, all using the unit mg /l or g/m^3:

- BOD_5-A, BOD_5-B, BOD_5-C, BOD_5-D, BOD_5-E (mg O_2 /l);
- NIT-A, NIT-B, NIT-C, NIT-D, NIT-E (mg N/l);
- AMM-A, AMM-B, AMM-C, AMM-D, AMM-E (mg N/l);
- TPO-A, TPO-B, TPO-C, TPO-D, TPO-E (mg P/l); and
- ORN-A, ORN-B, ORN-C, ORN-D, ORN-E (mg N/l).

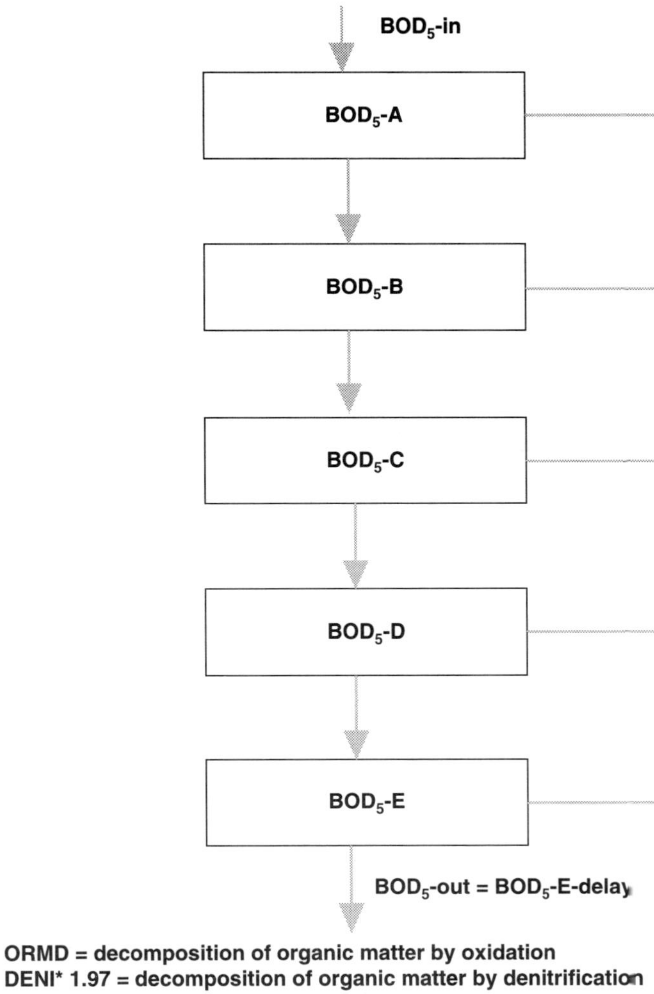

Figure 3.13: The BOD_5 submodel.

The model variables are expressed by three letters (e.g., NIT for nitrate), followed by IN, OUT or A, B, C, D or E, with the parameters using two letters.

The model has the following forcing functions, which the user must specify for a given model run:

- Volume of wetland (m^3; possible range 10–10,000,000);
- Flow of water (QIN, expressed as m^3/24 h; possible range 1–1,000,000);
- Porosity (as fraction of POR; no unit; range 0–1; default value 0.46);
- Input concentration of BOD_5 (BOD-IN; mg O2/l; range 0–1000);
- Input concentration of ammonium (AMM-IN; mg N/l; range 0–100);
- Input concentration of nitrate (NIT-IN; mg N/l; range 0–100);
- Input concentration of total phosphorus (TPO-IN; mg P/l; range 0–50);
- Concentration of organic nitrogen (ORN-IN; mg N/l; range 0–200)

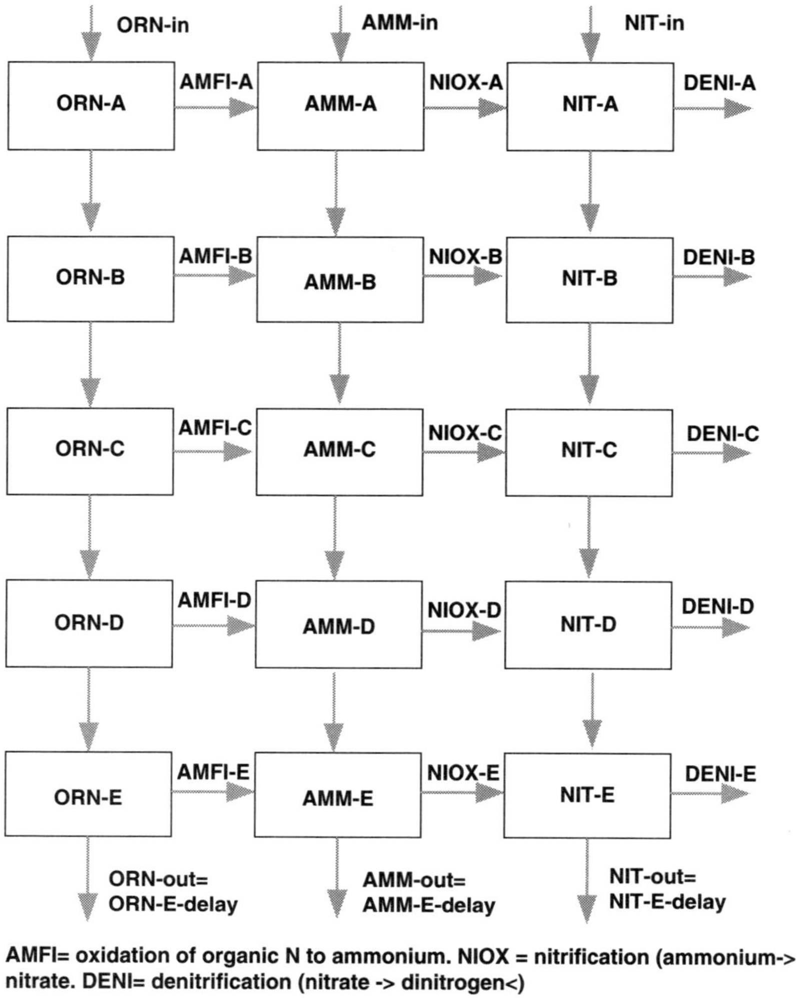

Figure 3.14: The nitrogen submodel, illustrating the three nitrogen compounds (organic-N, ammonium and nitrate).

- Fraction of BOD_5 as suspended matter (POM; no unit; range 0–1);
- Fraction of organic-N matter as suspended matter (PON; no unit);
- Fraction of phosphorus as suspended matter (POP; no unit);
- Average oxygen concentration in Box A (AOX; mg/l; range 0–20);
- Average oxygen concentration in Box B (BOX; mg/l; range 0–20);
- Average oxygen concentration in Box C (COX; mg/l; range 0–20);
- Average oxygen concentration in Box D (DOX; mg/l; range 0–20);
- Average oxygen concentration in Box E (EOX; mg/l; range 0–20);
- Default value for AOX, BOX, COX, DOX and EOX = 4 mg/l;
- Average temperature (TEMP; as function of time; daily average temperature is listed for the number of days to be simulated with the model);
- The length of model simulations must be indicated as number of days;

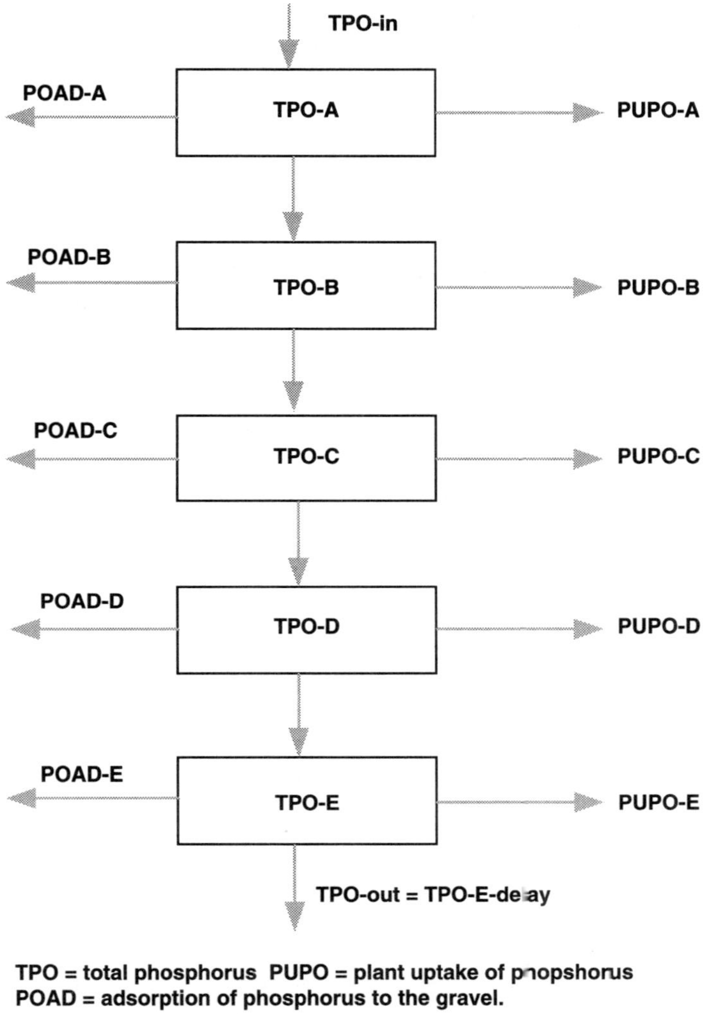

Figure 3.15: The phosphorus submodel.

- The following forcing functions are calculated, and included in the forcing function table, along with the forcing functions;
- Retention time, RTT (= VOL × POR/Q; 24 h);
- Retention time per box, RTB (= RTT/5; 24 h); and
- Box volume, BOV (= VOL × POR/5).

3.5.1 Process equations

A continuous transfer takes place from one state variable to another in the model simulations. This section identifies the processes that take place in the subsurface wetland. It is noted that the same processes take place in each box, although the concentrations are different in each box. Thus, the equations are repeated in the model program, with an indication of the concentrations in the 5 different model boxes – A, B, C, D and E. All the processes are expressed by four letters, followed by A, B, C, D or E, corresponding to the five boxes. It is reiterated that the model expressions are the same for each box, although the applied concentrations of the modelled materials will differ for each box. Exponent is expressed by the notation (^).

The following equations are repeated in each box, with an indication of the letters of the box (see also Figs 3.12–3.14):

- Ammonification = AMFI = ORN × AC × TA^(TEMP-20);
- Nitrification = NIOX = AMM × NC × INOX × TN^(TEMP-20)/(AMM + MA);
- Oxidation of BOD_5 = ORMD = BOD_5 × OC × INOO × TO^(TEMP-20);
- Denitrification = DENI = NIT × DC × TD^(TEMP-20)/(NIT + MN);
- INOX-A = AOX/(AOX + KO), and so on for boxes B, C, D and E, using the notations BOX, COX, DOX and EOX; however, KO is the same parameter for all five boxes;
- INOO-A = AOX/(AOX + OO), and so on for boxes B, C, D and E, using the notations BOX, COX, DOX and EOX; however, OO is the same parameter for all five boxes;
- Plant uptake of ammonium = PUAM = AMM × PA;
- Plant uptake of nitrate = PUNI = NIT × PN;
- Plant uptake of phosphorus = PUPO-A = TPO-A × PP × (1 – POP) for box A, while PUPO-B = TPO-B × PP, PUPO-C = TPO-C × PP, PUPO-D = TPO-D × PP and PUPO-E = TPO-E × PP (note that the multiplication by (1 – POP) only applied to box A); and
- Adsorption of phosphorus = POAD-A = TPO-A × (1 – POP) × (POR) – AF × (1 – POR), if POAD > 0; otherwise POAD = 0 for box A, while the following equation is applied for the other boxes: POAD = TPO × POR – AF × (1 – POR).

The model uses delay values (i.e., the concentrations of the five constituents in the five boxes during one box-retention time (RTB) later). For example,

AMM-A-delay = AMM-A at time t – RTB,
when t = or > RTB; if t < RTB, AMM-A-delay is 0.

These equations are repeated for all five constituents in all five boxes and the delay concentrations are indicated with "-delay".

Further, the simulated results also are used to determine the removal efficiencies, which are also shown on graphs. They are found as function of time, as follows:

- Efficiency of BOD_5-removal (%) = 100 × (BOD_5-in – BOD_5-out)/BOD_5-in;
- Efficiency of nitrate removal (%) = 100 × (NIT-in – NIT-out)/NIT-in;
- Efficiency of ammonium removal (%) = 100 × (AMM-in – AMM-out)/AMM-in;
- Efficiency of organic-N removal (%) = 100 × (ORN-in – ORN-out)/ORN-in;
- Efficiency of nitrogen removal (%) = 100 × ((NIT-in + AMM-in + ORN-in) – (NIT-out + AMM-out + ORN-out))/(NIT-in + AMM-in + ORN-in); and
- Efficiency of phosphorus removal (%) = 100 × (TPO-in – TPO-out)/TPO-in.

3.5.2 Parameters

The model parameters are as follows:

- AC = 0.05–0.8 [default value 0.5 (1/24 h)];
- NC = 0.1–1.5 [default value 0.8 (1/24 h)];

- OC = 0.05–0.8 [default value 0.5 (1/24 h)];
- DC = 0.25–5 [default value 2.2 (1/24 h)];
- TA = 1.02–1.06 [default value 1.04 (no unit)];
- TN = 1.02–1.07 [default value 1.047 (no unit)];
- TO = 1.02–1.06 [default value 1.04 (no unit)];
- TD = 1.05–1.12 [default value 1.09 (no unit)];
- KO = 0.1–2 [default value 1.3 (mg/l)];
- OO = 0.1–2 [default value 1.3 (mg/l)];
- MA = 0.05–2 [default value 1 (mg/l)];
- MN = 0.01–1 [default value 0.1 (mg/l)];
- PA = 0.00–1 [default value 0.01 (1/24 h)];
- PN = 0.00–1 [default value 0.01 (1/24 h)];
- PP = 0.00–1 [default value 0.003 (1/24 h)]; and
- AF = 0–100 [default value 1.0].

3.5.3 Differential equations

The 25 differential equations in the model are as follows:

- BOV × d BOD-5-A/dt = QIN × BOD-IN – QIN × (1 – POM) × BOD-A-delay – BOV × ORMD-A – DENi-A × 1.97;
- BOV × d BOD-5-B/dt = QIN × (1 – POM) × BOD-A-delay – BOV × ORMD-B – QIN × BOD-B-delay – DENI-B × 1.97;
- BOV × d BOD-5-C/dt = QIN × BOD-B-delay – BOV × ORMD-C – QIN × BOD-C-delay – DENI-C × 1.97;
- BOV × d BOD-5-D/dt = QIN × BOD-C-delay – BOV × ORMD-D – QIN × BOD-D-delay – DENI-D × 1.97;
- BOV × d BOD-5-E/dt = QIN × BOD-D-delay – BOV × ORMD-E – QIN × BOD-E-delay – DENI-E × 1.97
  (QIN × BOD-E-delay indicates BOD_5-OUT, which is eventually shown on a graph, together with measured values of the BOD_5-OUT, while BOD_5-A, B, C, D and E are shown in a table as function of time);
- BOV × dNIT-A/dt = QIN × NIT-IN – QIN × NIT-A-delay – BOV × DENI-A + BOV × NIOX-A – BOV × PUNI-A;
- BOV × dNIT-B/dt = QIN × NIT-A-delay – BOV × DENI-B + BOV × NIOX-B – BOV × PUNI-B – QIN × NIT-B-delay;
- BOV × dNIT-C/dt = QIN × NIT-B-delay – BOV × DENI-C – BOV × NIOX-C – BOV × PUNI-C – QIN × NIT-C-delay;
- BOV × dNIT-D/dt = QIN × NIT-C-delay – BOV × DENI-D + BOV × NIOX-D – BOV × PUNI-D – QIN × NIT-D-delay;
- BOV × dNIT-E/dt = QIN × NIT-D-delay - BOV × DENI-E – BOV × NIOX-E - BOV × PUNI-E – QIN × NIT-E-delay
  (QIN × NIT-E-delay indicates NIT-OUT, which is eventually shown on a graph, together with measured values of NIT-OUT, while NIT-A, B, C, D and E are all shown in a table as function of time);
- BOV × dAMM-A/dt = QIN × AMM-IN – QIN × AMM-A-delay – BOV × NIOX-A + BOV × AMFI-A – BOV × PUAM-A;

- BOV × dAMM-B/dt = QIN × AMM-A-delay + BOV × AMFI-B – BOV × NIOX-B – BOV × PUAM-B – QIN × AMM-B-delay;
- BOV × dAMM-C/dt = QIN × AMM-B-delay + BOV × AMFI-C – BOV × NIOX-C – BOV × PUAM-C – QIN × AMM-C-delay;
- BOV × dAMM-D/dt = QIN × AMM-C-delay + BOV × AMFI-D – BOV × NIOX-D – BOV × PUAM-D – QIN × AMM-D-delay;
- BOV × dAMM-E/dt = QIN × AMM-D-delay + BOV × AMFI-E – BOV × NIOX-E – BOV × PUNI-E – QIN × AMM-E-delay
  (AMM-E-delay indicates AMM-OUT, which is eventually shown on a graph, together with measured values of AMM-OUT, while AMM-A, B, C, D and E are all shown in a table as function of time);
- BOV × dORN-A/dt = QIN × ORN-IN – QIN × (1–PON) × ORN-A-delay – BOV × AMFI-A;
- BOV × dORN-B/dt = QIN × ORN-A-delay – BOV × AMFI-B - QIN × ORN-B-delay;
- BOV × dORN-C/dt = QIN × ORN-B-delay – BOV × AMFI-C – QIN × ORN-C-delay;
- BOV × dORN-D/dt = QIN × ORN-C-delay – BOV × AMFI-D– QIN × ORN-D-delay;
- BOV × dORN-E/dt = QIN × ORN-D-delay – BOV × AMFI-E – QIN × ORN-E-delay
  (QIN × ORN-E-delay indicates ORN-OUT, which is eventually shown on a graph, together with measured values of ORN-OUT, while ORN-A, B, C, D and E are all shown in a table as function of time);
- BOV × dTPO-A/dt = QIN × TPO-IN – QIN × (1 – POP) × TPO-A-delay – BOV × PUPO-A – BOV × POAD-A;
- BOV × dTPO-B/dt = QIN × (1 – POP) × TPO-A-delay – BOV × PUPO-B – BOV × POAD-B – QIN × TPO-B-delay;
- BOV × dTPO-C/dt = QIN × TPO-B-delay – BOV × PUPO-C – BOV × POAD-C – QIN × TPO-C-delay;
- BOV × dTPO-D/dt = QIN × TPO-C-delay – BOV × PUPO-D – BOV × POAD-D – QIN × TPO-D-delay; and
- BOV × dTPO-E/dt = QIN × TPO-D-delay – BOV × PUPO-E – BOV × POAD-E – QIN × TPO-E-delay
  (QIN × TPO-E-delay indicates TPO-OUT, which is eventually shown on a graph, together with measured values of TPO-OUT, while TPO-A, B, C, D and E are all shown in a table as function of time).

3.5.4 Model results

As mentioned above, the simulated values of BOD_5-out, nitrate-out (NIT-out), ammonium-out (AMM-out), total phosphorus-out (TPO-out) and organic nitrogen-out (ORN-out) are shown in the form of both tables and as graphs. If the measured values are available, they are shown on the same graphs, as a means of allowing a direct comparison.

The simulated results of the removal efficiencies are also shown on graphs, including:

- Efficiency of BOD_5-removal (%);
- Efficiency of nitrate removal (%);
- Efficiency of ammonium removal (%);
- Efficiency of organic-N removal (%);
- Efficiency of nitrogen removal (%);
- Efficiency of phosphorus removal (%).

It also may be useful to include the predicted concentrations of the five constituents in the five boxes, as a means of illustrating where the removal processes are most effective in the wetland, and where they are less effective. It may be possible to apply such information to improve the overall removal efficiencies by imposing changes in the composition of the wastewater, or by changes directly in the wetland (e.g., addition of oxygen). The predicted concentrations in the boxes obtained with the model simulations are listed in a table for each day in the simulation period, as follows: BOD_5-A, BOD_5-B, BOD_5-C, BOD_5-D, BOD_5-E, NIT-A, NIT-B, NIT-C, NIT-D, NIT-E, AMM-A, AMM-B, AMM-C, AMM-D, AMM-E, TPO-A, TPO-B, TPO-C, TPO-D, TPO-E, ORN-A, ORN-B, ORN-C, ORN-D and ORN-E.

3.5.5 Practical information about forcing functions and parameters

Usually the model is applied to design a wetland that has not yet been constructed. Thus, it is recommended that the following information is obtained, in order to be able to indicate the forcing functions:

1. Measure the temperature every 6 hours, for example, in a shallow aquatic ecosystem (i.e., before the wetland is constructed, and directly in the wetland after it has been constructed) to get an initial indication of the temperature variations. Alternatively, use some other local temperature measurements. Find the daily average temperature to be listed on the forcing function screen image. Although the temperature expression is not linear, using the average daily temperature for the temperature expression applied herein is only a minor error.
2. Use the default values indicated, when no information about the forcing functions is available.
3. For the uptake rate coefficients of nutrients (nitrate, ammonium and phosphorus; PA, PN and PP), use knowledge about the phosphorus (P) and nitrogen (N) content, and the growth rate of plants that will be applied in the wetland.

3.5.6 Use of the model for wetland design

If the model is used to design a wetland that has not yet been constructed, the concentrations in the wetland, the water flow through the wetland and the expected concentrations after wetland treatment, should be known. The model is then used to predict the results for simulating different volumes. The simulated results that meet the criteria for the treated water are used, and the corresponding volume is used for the design. It is recommended that a volume 10–15% greater than that predicted with this method be used, in order to take into account the uncertainties of the model. Further, a depth of 0.6–1.00 m is used, and a maximum flow rate of about 1.25 m/h applied, in order to determine the width and length of the wetland.

3.6 River models

The main environmental problem of lakes and reservoirs is eutrophication, resulting from a too high concentration of nutrients. The core questions are:

1. What is the concentration of the limiting nutrient, which is often phosphorus?
2. What is the primary production?
3. What is the transparency, at the time when the eutrophication is at maximum?

For rivers, the main environmental problem is the oxygen balance. Discharge of organic matter causes oxygen depletion. The core problems are:

1. What is the concentration of organic matter, expressed frequently as BOD_5 mg/l, that is, the amount of oxygen that the decomposition of the organic matter during a period of 5 days will consume?
2. What is the oxygen concentration, which is function of the distance from the discharge point of organic matter? What is the minimum oxygen concentration?

Below is presented a river model that is able to answer the questions 1 and 2. The model is presented by the use of the software STELLA, which was briefly mentioned in Chapter 1. The example should enable the readers easily to develop their own river model by using STELLA. After presentation of a simple river model, it will be discussed which processes it probably would be beneficial to include, if the model should be expanded to include more components and interactions. It is, however, necessary to introduce the basic equations to uncover the basic ideas behind river models: for the readers.

Organic matter is decomposed with good approximations by a first order reaction. If L is the concentration of organic matter (mg/l) and k_1 is the rate coefficient for the decomposition, the following differential equation is valid:

$$dL/dt = -k_1L \tag{3.19}$$

The equation has the following analytical solution:

$$L_t = L_0 \exp(-k_1t) \tag{3.20}$$

where L_t is the concentration at time t and L_0 is the initial concentration.

L may be expressed as BOD_5 – it means as mg/l oxygen consumption during a period of 5 days. If it is expressed as mg/l (average) organic matter or detritus, the concentration has accordingly to the chemical decomposition equation to be multiplied by 1.39. One gram detritus or organic matter requires, in other words, 1.39 g oxygen to be decomposed as much as it is possible during a period of 5 days (which is close to 100% decomposition).

Nitrification of ammonium is also causing oxygen depletion. If the ammonium concentration is denoted as NC (mg N/l in form of ammonium) and it is presumed that nitrification follows a first order reaction, the following differential equation is valid:

$$dNC/dt =- k_NNC \tag{3.21}$$

where k_N is the rate coefficient for the nitrification. Equation (3.21) has the following solution:

$$NC_t = NC_0 \exp(-k_Nt) \tag{3.22}$$

where NC_t is the concentration at time t and NC_0 is the initial concentration.

Notice that NC is the concentration in mg ammonium-N/l and the corresponding oxygen consumption is found from the chemical equation, for the nitrification:

$$NH_4^+ + 2O_2 \rightarrow NO_3^- + H_2O + H^+ \tag{3.23}$$

It means that 1 g of ammonium-N requires 2 × 32/14 = 4.6 g of oxygen, which will be included in the model, when the nitrification is "translated" to oxygen depletion. The factor is, however, in practice 4.3 not 4.6, due to the bacterial assimilation of ammonium by the nitrifying micro-organisms.

It could be considered to apply a Michaelis–Menten equation instead of a first order expression for eqns (3.19) and (3.21) [1]. According to the Michaelis–Menten expressions used in eutrophication models (see Section 3.3), eqn (3.19) could be multiplied by:

$$[Ox]/(k_{mo} + [Ox]) \tag{3.24}$$

to account for the influence of oxygen as a limiting factor of the decomposition rate. Similarly, eqn (3.21) could be multiplied by

$$\min\left([NC]/(k_{ma} + [NC]),\ [Ox]/(k_{ao} + [Ox])\right) \tag{3.25}$$

to account for the influence of both ammonium and oxygen as possible limiting factors of the decomposition. Below, when the model has to be created, the Michaelis–Menten expressions will be applied, but it is also possible to get reasonable good results by use of the first order expressions, which have the advantage that they can be solved analytically.

The decomposition of organic matter and nitrification are of course temperature dependent. A simple Arrhenius expression may be applied:

$$\begin{aligned}&\text{The rate coefficient at temperature (°C) T}\\ &= \text{rate coefficient at 20°C} \times K^{(T-20)}\end{aligned} \tag{3.26}$$

K is with good approximation 1.05 for decomposition of organic matter, while the nitrification is more sensitive to temperature changes and K is therefore for this process 1.07–1.08 [4].

Typical values for the rate coefficients and the initial concentrations for various sources of organic matter and ammonium are shown in Table 3.5.

If the oxygen concentration is below the saturation concentration, that is, that the water is in equilibrium with the atmosphere, re-aeration from the atmosphere takes place. The equilibrium concentration can be found by Henry's Law. The saturation concentration is dependent on the temperature and the salinity of the water. In Appendix 1, Table 1, the equilibrium concentration of oxygen can be found as function of the temperature and the salinity.

Table 3.5: Characteristic values for k_1, k_N (1/24 h) and initial concentrations (mg/l) for various sources to oxygen depletion in streams and rivers.

Source	k_1	k_N	L_0	NC_0
Municipal wastewater	0.35–0.40	0.15–0.25	180–300	20–45
Mechanically treated w.w.	0.32–0.36	0.10–0.15	100–200	18–35
Biologically treated w.w.	0.10–0.25	0.05–0.20	10–40	15–32
Potable water	0.05–0.10	0.03–0.06	0–2	0–1
River water (averagely)	0.05–0.15	0.04–0.10	1–4	0–2
Agricultural drainage water	0.08–0.20	0.04–0.12	5–25	0–10
Wastewater, food industry	0.4–0.5	0.1–0.25	200–5000	20–200

The aeration is proportional to the difference between the oxygen concentration at saturation, Ox_{sat} and the actual oxygen concentration, [Ox]. The driving force for the aeration is this difference. It is expressed by the following equation:

$$\text{Re-aeration} = d[Ox]/dt = K_a\,(Ox_{sat} - [Ox]) \quad (3.27)$$

where K_a is the re-aeration coefficient (1/24 h).

K_a is dependent on the temperature and the flow rate of water. The aeration is finally also proportional to the surface area relative to the volume – it means it is inversely proportional to the water depth. There are several hundred empirical equations that can be used to estimate the re-aeration or the re-aeration coefficient. An often applied equation is:

$$K_a = 2.26 \times v \times \exp\,(0.024 \times (T - 20)/d, \quad (3.28)$$

Where v is the water flow rate in m/s, T is the temperature in degree Celsius and d is the depth of the stream or river in m.

The oxygen concentration is determined by the difference between the consumption and the re-aeration. If we only consider decomposition of organic matter and nitrification, the oxygen concentration is determined by the following differential equation:

$$d[Ox]/dt = K_a\,(Ox_{sat} - [Ox]) - k_1L - k_N NC \quad (3.29)$$

The solution of the differential equations for L and NC can be used in this differential equation to yield the following expression for the solution of [Ox] as function of time:

$$d[Ox]/dt = K_a\,(Ox_{sat} - [Ox]) - k_1L_0 \exp(-k_1t) - k_N NC_0 \exp(-k_Nt) \quad (3.30)$$

This equation can be solved analytically, but below we have used STELLA to determine [Ox], L and NC as function of time in accordance with the shown equations. It is of course possible to add many more processes to the model such as primary production by phytoplankton and macrophytes producing oxygen, denitrification consuming organic matter as carbon source, presence of organic matter with different biodegradability, changed re-aeration at turbulent flow and so on. As always, when we develop models, the problem, the available data and the system with a wide spectrum of possible processes should determine the complexity of the model. Let us, however, develop the model by use of the software STELLA to illustrate how to use this software and to illustrate how powerful even relatively simple models are.

The STELLA diagram is shown in Fig. 3.16. Notice that the three state variables – organic matter (L), ammonium–nitrogen (NC) and oxygen (Ox) are covered by three different submodels that are connected. The oxygen consumption is due to the decomposition of organic matter and the nitrification, and the oxygen concentration is influencing the decomposition rate and nitrification as presented in eqns (3.24) and (3.25). Time is considered as the independent variable, but it could also be the distance from the discharge, for instance, of wastewater. If the water flow rate, for instance, is 0.1 m/s, the 24 hours will correspond to $0.1 \times 3600 \times 24\,m = 8640\,m$. The model has a constant discharge, which can be considered to be agricultural drainage water along the shoreline of the stream. The point discharge of wastewater takes place at time 0 and will therefore correspond to the initial value of L and NC. The dilution has, of course, to be considered when the initial values are calculated. If for example, $500\,m^3/h$ of wastewater with 30 mg/l of BOD_5 and 12 mg/l of ammonium-N is discharged into a river with $4500\,m^3/h$ of waterflow,

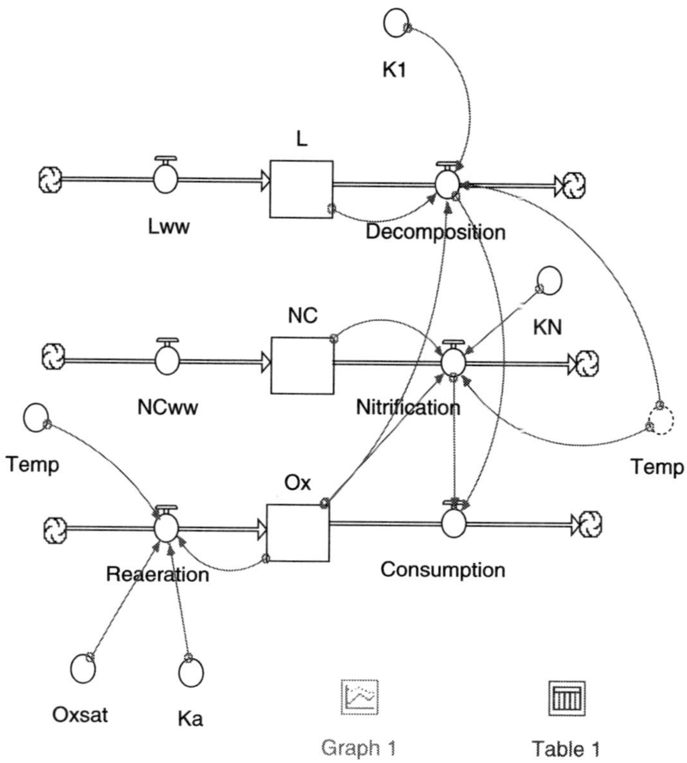

Figure 3.16: Conceptual diagram of the presented river model by application of the software STELLA.

then the dilution factor is 10. If the river water has 5 mg BOD_5/l and 2 mg ammonium-N/l, the mixture of river water and wastewater will have (5 × 9 + 30 × 1)/10 = 7.5 mg BOD_5/l and (2 × 9 + 12 × 1)/10 = 3 mg ammonium-N/l, which are applied as the initial concentrations. The model should be able to give information about how these concentrations are changed over time. The concentration of oxygen in treated wastewater will almost always be close to 0 mg/l. If the river water has 8 mg/l oxygen, the mixture of wastewater and river water will have an oxygen concentration of 7.2 mg/l corresponding to a dilution of the wastewater by a factor 10.

Figure 3.17 shows the result of using the following initial values:

- for BOD_5, 7.5 mg/l;
- for ammonium-N, 3 mg/l and
- for oxygen, 7.2 mg/l.

The result of running the model 90 days is shown in Fig. 3.17 and the equations are presented in Fig. 3.18. Notice the form of the differential equations that are applied in STELLA. Time can be translated to distance from the discharge point by the flow rate. If the cross sectional area is 50m^2, the water flow of 5000 m^3/h corresponds to 100 m/h. Twenty four hours

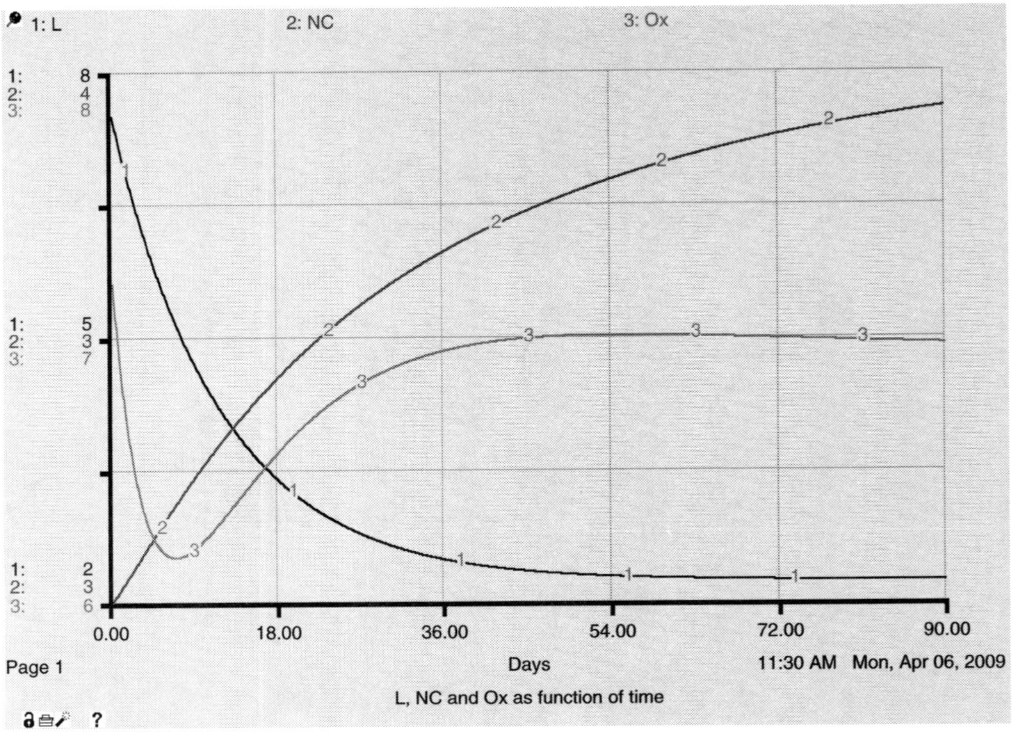

Figure 3.17: Results of the river model. Oxygen has minimum 6.2 mg oxygen/l after 8 days. The ammonium–nitrogen continues to increase inspite of the nitrification due to the input of ammonium-N from non-point sources, while the BOD_5 is decreasing corresponding to decomposition. Only a small amount of BOD_5 is coming from non-point sources. In accordance with the flow rate, the minimum concentration occurs after 19.2 km.

will therefore correspond to 2400 m and the 90 days to 216 km. The minimum oxygen concentration occurs after 8 days or 19.2 km. Table 3.6 gives the model results in a tabular form for every 5 days.

Illustration 3.3

How would the results of the model presented above in Table 3.6 and Figs 3.16–3.18 be changed if:

(a) untreated wastewater with a BOD_5 of 155 mg/l would be discharged?
(b) the organic matter would decompose 2.5 times faster?
(c) untreated wastewater with a BOD_5 of 155 mg/l would be discharged and the organic matter would decompose 2.5 times faster?
(d) the discharge of ammonium-N from non-point sources – mainly agricultural drainage water – would be ten times more concentrated. Would the nitrification of this much higher amount of ammonium give significant higher oxygen depletion? It is presumed that the nitrification rate coefficient is at the same time doubled to take "the worst situation" into account. In other words: could the agricultural drainage water have a major influence on the oxygen balance for streams and rivers?

```
L(t) = L (t – dt) + (Lww – Decomposition) × dt
INIT L = 7.5

INFLOWS:
Lww = 0.2
OUTFLOWS:
Decomposition = (L × K1 × Ox × 1.05^(20 –Temp)/(Ox + 2.5))
NC(t) = NC (t – dt) + (NCww – Nitrification) × dt
INIT NC = 3

INFLOWS:
NCww = 0.1
OUTFLOWS:
Nitrification = NC × KN × MIN (Ox/(Ox + 3), NC/(NC + 1)) × 1.075^(Temp – 20)
Ox(t) = Ox (t – dt) + (Re-aeration – Consumption) × dt
INIT Ox = 7.2

INFLOWS:
Re-aeration = Ka × (Oxsat–Ox) × exp (0.024 × (Temp – 20))
OUTFLOWS:
Consumption = Decomposition + 4.3 × Nitrification
K1 = 0.1
Ka = 0.226
KN = 0.05
Oxsat = 10
Temp = 16
```

Figure 3.18: The equations of the river model. Notice that STELLA is using differential equations, which are based on the same template as for differential equations, namely: rate of accumulation = inputs – outputs.

Table 3.6: Model results in a tabular form. Concentrations (BOD_5 or L, NC or ammonium-N and oxygen) are all in mg/l and process rates are in mg/l per 24 h.

Day	BOD_5	NC	Oxygen	Ox consumption	Decomp.	Nitrification
0	7.50	3.00	7.20	1.01	0.67	0.08
5.0	5.64	3.11	6.21	0.83	0.49	0.08
10.0	4.47	3.21	6.24	0.74	0.39	0.08
15.0	3.70	3.30	6.45	0.69	0.32	0.08
20.0	3.19	3.37	6.64	0.66	0.28	0.09
25.0	2.86	3.43	6.78	0.64	0.25	0.09
30.0	2.64	3.48	6.88	0.63	0.24	0.09
35.0	2.50	3.52	6.94	0.62	0.22	0.09
40.0	2.40	3.56	6.98	0.62	0.21	0.09
45.0	2.34	3.59	7.00	0.61	0.21	0.09
50.0	2.30	3.62	7.00	0.61	0.21	0.09
55.0	2.28	3.64	7.01	0.61	0.20	0.10
60.0	2.26	3.66	7.01	0.62	0.20	0.10
65.0	2.25	3.68	7.00	0.62	0.20	0.10
70.0	2.24	3.70	7.00	0.62	0.20	0.10
75.0	2.24	3.71	6.99	0.62	0.20	0.10
80.0	2.24	3.73	6.99	0.62	0.20	0.10
85.0	2.24	3.74	6.98	0.62	0.20	0.10
Final	2.24	3.75	6.98			

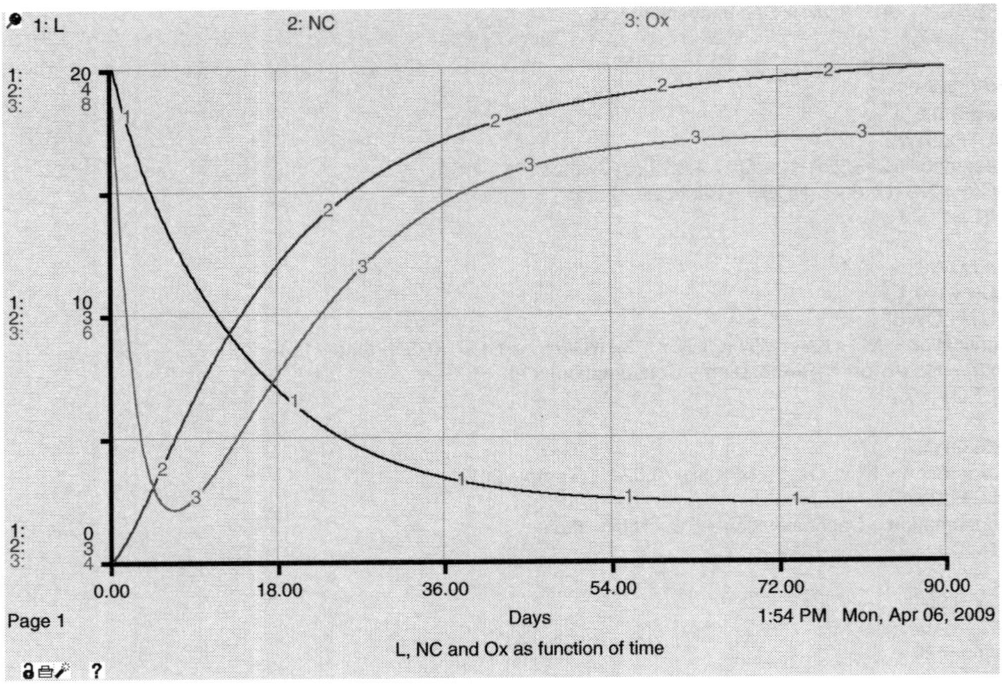

Figure 3.19: The initial value of BOD_5 is changed to be 20 mg/l. Compared with 3.15, it is seen that the minimum oxygen concentration occurs at about the same time, but now it is about 4.4 mg oxygen/l only – entirely due to the higher initial value.

Solution

(a) The model is run by an initial value of $BOD_5 = (9 \times 5 + 155)/10 = 20$ mg/l.
(b) The model is run by a decomposition rate coefficient, K1 = 0.25 1/24 h.
(c) The model is run by an initial value of $BOD_5 = (9 \times 5 + 155)/10 = 20$ mg/l and a decomposition rate coefficient, K1 = 0.25 1/24 h.
(d) The ammonium–N originated from non-point sources, mainly agricultural drainage water, is increased from addition of 0.1 mg/l per 24 h to 1 mg/l per 24 h and the nitrification rate is at the same time changed from 0.05 to 0.1 1/24 h.

The results of the four simulations are shown in Figs 3.19–3.22. The results show that the initial concentration of organic matter has mainly an influence on the minimum oxygen concentration, while the decomposition rate coefficient has influence also on the time when the minimum oxygen concentration would occur. A 10 times higher ammonium-N from agricultural drainage water would change the oxygen balance significantly. The oxygen concentration would almost disappear and the re-aeration would not be sufficient to compensate for the oxygen consumption. There would, therefore, be no minimum oxygen concentration but there would be a steady decrease. Due to formation of almost anaerobic conditions, the decomposition rate of organic matter is decreased, which is taken into account by the Michaelis–Menten expression (see eqn (3.22)) where the rate is dependent on the oxygen concentration.

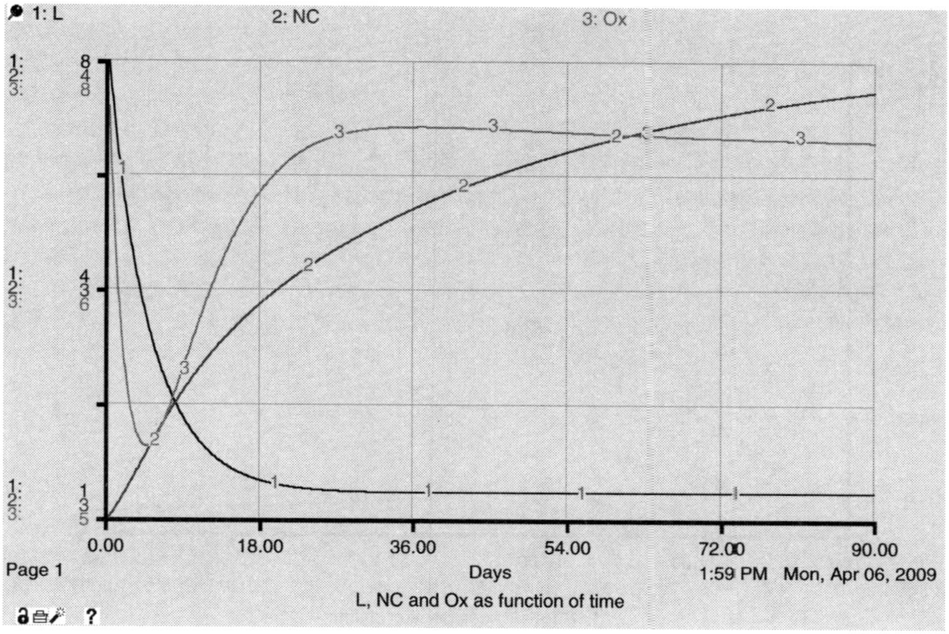

Figure 3.20: The model is run by a decomposition rate coefficient, $K_1 = 0.25$ 1/24 h. Now the minimum oxygen concentration occurs earlier, namely after about 5 days. The minimum concentration is also lower – about 5.3 mg/l than in Fig. 3.16.

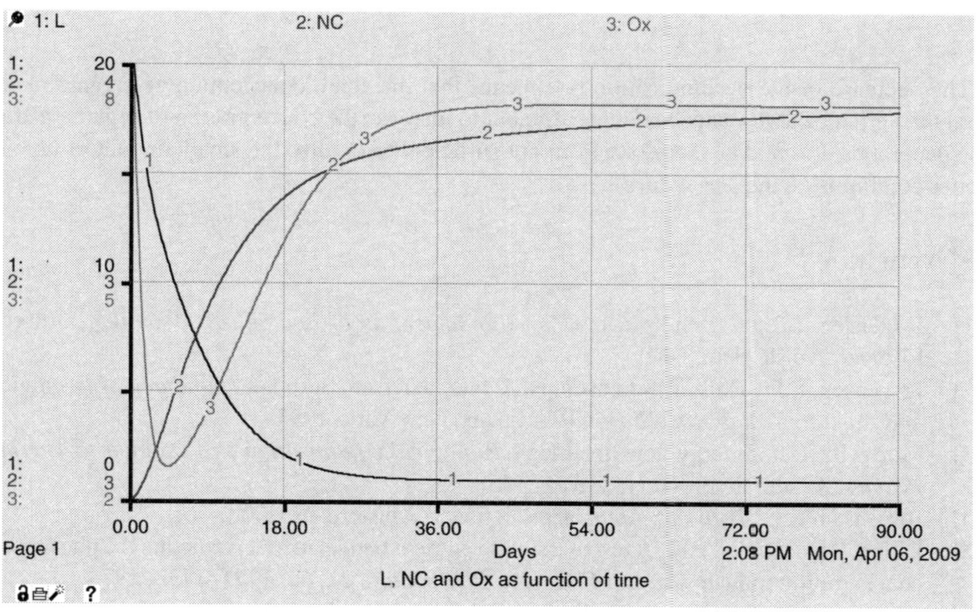

Figure 3.21: The initial value of $BOD_5 = 20$ mg/l and the decomposition rate coefficient, $K_1 = 0.25$ 1/24 h. The minimum oxygen concentration is about 2.4 mg/l and it occurs after 5 days.

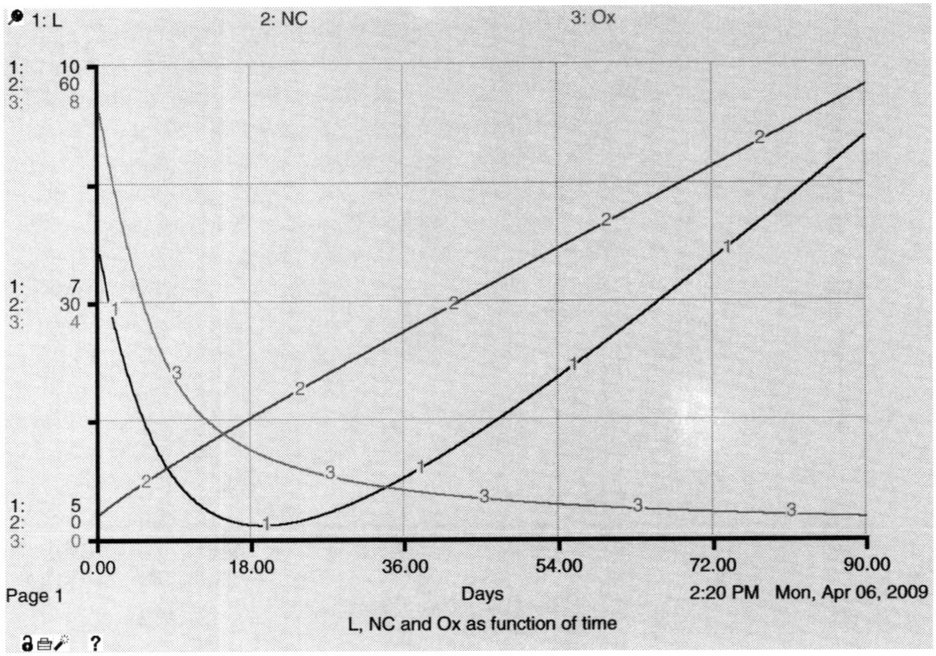

Figure 3.22: The ammonium–N originating from non-point sources, mainly agricultural drainage water, is increased from addition of 0.1 mg/l per 24 h to 1 mg/l per 24 h and the nitrification rate is at the same time changed from 0.05 to 0.1 1/24 h.

The decrease of the decomposition rate means that the small contribution of organic matter from the agricultural drainage water is sufficient to increase the concentration of organic matter. The decomposition rate is, therefore, even not sufficient to oxidize the small amount of organic matter coming from the agriculture.

References

[1] Jørgensen, S.E. & Bendoricchio, G., *Fundamentals of Ecological Modelling*, 3rd edn, Elsevier: Amsterdam, 2001.

[2] Jørgensen, S.E., Halling-Sørensen, B. & Nors Nielsen, S. (eds), *Handbook of Ecological and Environmental Modelling*, CRC: Lewis, New York, 1995.

[3] Morowitz, H.J., Energy flow in biology. *Biological Organisation as a Problem in Thermal Physics*, Academic Press: New York, 1968.

[4] Jørgensen, S.E., *Pollution Abatement*, Elsevier: Amsterdam, 2000.

[5] Dillon, P.J. & Rigler, F.H., A test of a simple nutrient budget model predicting the phosphorus concentration in lake water. *J. Fisk. Res. Board Can.*, **31**, pp. 1771–1778, 1974.

[6] Jørgensen, S.E., An eutrophication model for a lake. *Ecol. Modelling*, **2**, pp. 147–166, 1976.

[7] Jørgensen, S.E., Examination of the generality of eutrophication models. *Ecol. Modelling*, **32**, pp. 251–266, 1986.

[8] Jørgensen, S.E., Validation of "Prognosis based upon an eutrophication model". *Ecol. Modelling*, **32**, pp. 165–182, 1986.

[9] Jørgensen, S.E., Structural dynamic model. *Ecol. Modelling*, **31**, pp. 1–9, 1986.

[10] Di Toro, D.M. & Conolly, J.F., Jr., Mathematical models of water quality in large lakes, Part II: Lake Erie. U.S. Environmental Protection Agency, Ecological Research Series, EPA-600/3-80-065, 1980.

[11] Jørgensen, S.E., Löffler, H., Rast, W. & Straskraba, M., *Lake and Reservoir Management*, Elsevier: Amsterdam, 2004.

[12] Jørgensen, S.E., Chon, T. & Recknagel, F., *Handbook of Ecological Modelling and Informatics*, WIT Press: Southampton, 2009.

Exercises/problems – Chapter 3

1. BOD_5 at 20°C is 100 mg/l. What is BOD_7 at 15°C?
2. A completely mixed lake has an inflow of 40 l/s, a depth of 4 m and an area of 20 ha. In the inflowing water there is a toxic substance at a concentration of 50 mg/l. The biological half-life for the toxic substance is 5 days. The toxic substance is soluble in water and it has a high boiling point. What is the expected concentration of the toxic substance in the lake?
3. Estimate the chlorophyll a concentration and the transparency for a lake with a total concentration of 0.45 mg P/l and 3.5 mg N/l.
4. Why is the validation of a developed model almost always compulsory?
5. Explain why the generality of a eutrophication model is decreasing with increasing model complexity?
6. The settling of phytoplankton and suspended matter is dependent on the temperature. Explain how the dependence could be considered in the model?
7. If top-carnivorous fish would be important to include in a eutrophication model, which equations would you propose to use? Under which circumstances would it be necessary to include top-carnivorous fish in eutrophication models?
8. The subsurface wetland model presented in Section 3.5 has divided the wetland into five boxes. Give the advantages and disadvantages by increasing the number of boxes to 7, 8 or 10. How much would you presume it would influence the analytical results of the treated water to include more boxes? Do you find that the proposed five boxes is a suitable number?
9. A subsurface wetland has the following analysis for the treated water: BOD_5 32.0 mg/l, ammonium-N 5.1 mg/l, nitrate-N 12.3 mg/l, organic-N 12.7 mg/l and total P 6.5 mg/l. Indicate if the five concentrations would increase or decrease or remain unchanged if air could penetrate deeper into the wetland, that is, the oxygen concentration in the root zone would be higher. Indicate that it would mean for the analytical results of the treated water.
10. By heavy rain the water in a subsurface wetland will be diluted. If it is presumed that precipitation dilutes the water by a factor 2, what would be the effect on the analytical data of the treated water and thereby on the loading on the receiving water? Explain why it would be important in many cases to consider the dilution by rain water?
11. In Sweden BOD_7 is applied instead of BOD_5. Indicate the advantages of using 7 days instead of 5 days for determination of the biological oxygen demand. How would it influence the results of the presented river models?
12. The organic matter concentration in a wastewater sample is 150 mg/l, and it is with good approximation corresponding to an oxygen demand in accordance with the stoichiometry of the decomposition process that is 1.39 times the concentration of organic matter. What is BOD – infinite? Set up a graph that is plotting BOD_x versus x, where x is the time in days, which is used to determine the biological oxygen demand for the wastewater sample. How many days are needed to determine 90%, 95% and 98% of the total oxygen demand corresponding to decomposition of the organic matter (the mentioned 150 mg/l)? The decomposition rate coefficient for wastewater may be applied. The temperature is 20°C.
13. BOD_5 is determined for a wastewater sample to be 180 mg/l. Find BOD_{10}, BOD_2 and BOD-infinite. The decomposition rate coefficient for wastewater may be applied. The temperature is 20°C.

14. The oxygen demand may include or may not include nitrification. Addition of methyl blue is usually applied to exclude nitrification. Find the difference in oxygen demand during a week (BOD_7) in percentage with and without nitrification for wastewater where the following information is valid:
    (a) 138 mg/l of organic matter demanding for a complete decomposition of 1.39 times as much oxygen.
    (b) The ammonium-N concentration is 32 mg/l.
    (c) The decomposition rate coefficients for organic matter and for nitrification for wastewater may be applied. The temperature is 20°C.

Comment on the results. Explain why it is most often necessary to get the BOD_5 both with and without nitrification included.

CHAPTER 4

Population dynamic models

4.1 Introduction

This chapter covers population models, where state variables are numbers or biomass of individuals or species. The growth of a population is considered (see Sections 4.2 and 4.3) with presentation of the basic concepts and equations. Afterward, the interactions between two or more populations are presented. The famous Lotka–Volterra model as well as several models that are more realistic is shown. Age distribution is introduced and computations by use of matrix models to cover the numbers in the age classes are illustrated. The two last sections of the chapter focus on metapopulation models and models of population infection.

4.2 Basic concepts

As early as in the 1920s, Lotka and Volterra developed the first population model, which is still widely used today [1, 2]. Since then many population models have been developed, tested and analyzed, and it will not be possible in this context to give a comprehensive review of these numerous models. The chapter will mainly focus on models of age distribution, growth and species interactions. Only deterministic models will be mentioned. Those who are interested in stochastic models can refer to Jørgensen *et al.* [3], which has a chapter devoted to stochastic, population dynamic models.

A population is defined as a collective group of organisms of the same species. Each population has several characteristic properties, such as population density (population size relative to available space), natality (birth rate), mortality (death rate), age distribution, spatial dispersion, growth forms and others.

A population is a currently changing entity and we are interested in its size and growth = change in size. If N represents the number of organisms and t the time, then dN/dt = the rate of change in the number of organisms per unit time at a particular instant (t) and dN/Ndt = the rate of change in the number of organisms per unit time per individual at a particular instant (t). If the population is plotted against time a straight line tangential to the curve at any point, that is, the currently change in slope of the curve represents the growth rate.

We may distinguish between absolute **natality** meaning the new individuals appearing per unit time: B_a = dN/dt and relative natality, which is the absolute natality divided by the number of individuals of the population: B_s = dN/Ndt. N indicates that only the new individuals are accounted for. **Mortality** refers to the death of individuals in the population. The absolute mortality rate, which is defined as the change of the population due to mortality: $M_a = dN/dt$ and the relative mortality, which is the absolute mortality relative to the population: $M_s = dN/Ndt$. N indicates that only the dead individuals are accounted for. The growth of population is the difference between absolute natality and absolute mortality:

$$\begin{aligned}\text{Growth} &= \text{natality} - \text{mortality}\\ &= dN/dt = dN/dt \text{ (due to natality)} - dN/dt \text{ (due to mortality)}\end{aligned} \quad (4.1)$$

Correspondingly, the relative growth = relative natality – relative mortality:

$$dN/Ndt = dN_n/Ndt - dN_m/Ndt \quad (4.2)$$

where N_n indicates that it is the population increase due to natality and N_m indicates that it is the population decrease due to mortality.

4.3 Growth models in population dynamics

The simplest growth models consider only one population. Its interactions with other populations are taken into consideration by the natality and the mortality, which might be considered dependent on the magnitude of the considered population but independent of other populations. In other words, we consider only one population as state variable.

The simplest growth model assumes *unlimited resources and exponential population growth.* A simple differential equation can be applied:

$$dN/dt = B_s \times N - M_s \times N = r \times N \quad (4.3)$$

where B_s is the relative natality or the instantaneous birth rate per individual, M_s is the relative mortality or the instantaneous death rate, $r = B_s - M_s$, N is here representing the number of individuals in the population or the biomass of the population and t the time. As seen, the equation represents first order kinetics (see Section 1.4 and Table 1.1) and exponential growth. If r is constant, then after integration we get:

$$N_t = N_0 \times e^{rt} \quad (4.4)$$

where N_t is the population density at time t and N_0 the population density at time 0.

Figure 4.1 represents a logarithmic graph of this equation. It is a straight line and r indicates the slope. If the logarithm of the number of individuals or the biomass versus the time is a straight line with a standard deviation that is in accordance with the uncertainty, it can be concluded that the growth is following the first order growth equation and that the slope of the line is equal to the growth rate.

The net reproductive rate (R_0) is defined as the average number of age class zero offspring produced by an average newborn organism during its entire lifetime.

The so-called **intrinsic rate of natural increase (r)** is dependent on the age distribution and it is only constant when the age distribution is stable. When R_0 is as high as possible, that is, under optimal conditions and with a stable age distribution, the maximal rate of natural increase is realized and designated r_{max}. Among various animals it ranges over several orders of magnitude (see Table 4.1).

Exponential growth is a simplification, which can only be valid over a certain time interval. Eventually every population must encounter the limitation of food, water, air or space, as the

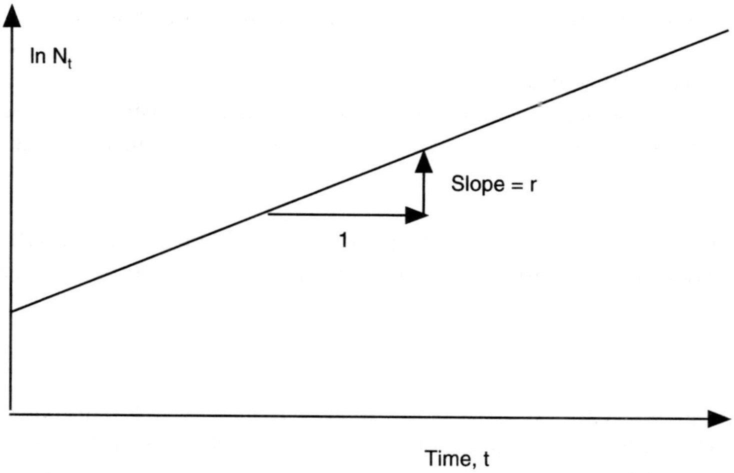

Figure 4.1: Logarithmic graph of the exponential growth equation.

Table 4.1: Estimated maximal instantaneous rate of increase (r_{max}, per capita per day) and mean generation time (in days) for a variety of organisms.

Taxon	Species	r_{max}	Generation time
Bacterium	*Escherichia coli*	ca. 60.0	0.014
Algae	*Scenedesmus*	1.5	0.3
Protozoa	*Paramecium aurelia*	1.24	0.33–0.50
Zooplankton	*Daphnia pulex*	0.25	0.8–2.5
Insect	*Tribolium confusum*	0.120	ca. 80
Insect	*Rhizopertha dominica*	0.085 (0.07–0.10)	ca. 100
Insect	*Ptinus sexpunctatus*	0.006	215
Octopus	various species	0.01	150
Mammal	*Rattus norwegicus*	0.015	150
Mammal	*Canis domesticus*	0.009	ca. 1000*
Mammal	*Homo sapiens*	0.0003	ca. 7000*
Sea turtle		0.0002	70,000

*Under natural conditions.

world is finite. To account for this, we apply the concept of density dependence, that is, vital rates (r) depend on population size (N) (while we now ignore differences caused by age). Let the **carrying capacity (K)** be defined as the density of organisms at which r is zero. At zero density, R_0 is maximal and r becomes r_{max}. The logistic growth equation considers the density dependence by use of the carrying capacity and has already been briefly mentioned in Chapter 1. The logistic equation is formulated mathematically as follows:

$$dN/dt = r \times N \times (1 - N/K) \tag{4.5}$$

where K is the carrying capacity.

The application of the logistic growth equation requires three assumptions:

1. That all individuals are equivalent.
2. That K and r are immutable constants independent of time, age distribution etc.
3. That there is no time lag in the response of the actual rate of increase per individual to changes in N.

All the three assumptions are unrealistic and can be strongly criticized. Nevertheless, several population phenomena can be properly illustrated by use of the logistic growth equation.

Illustration 4.1

An algal culture shows a carrying capacity due to the self-shading effect. In spite of "unlimited" nutrients, the maximum concentration of algae in a chemostat experiment was measured to be 125 g/m^3. At time 0, 0.1 g/m^3 of algae was introduced and after 2 days, a concentration of 1 g/m^3 was observed. Set up a logistic growth equation for these observations.

Solution
During the first 5 days, we are far from the carrying capacity and we have with good approximations:

$$\ln 10 = r_{max} \times 2$$

$$r_{max} = 1.2 \text{ day}^{-1}$$

and since the carrying capacity is 125 g/m^3, we have: (C = algae concentration)

$$dC/dt = 1.2 \times C \times (125 - C/125)$$

Integration and use of the initial condition C (0) = 0.1 yield

$$C = 125/(1 + e^{(a-1.2\,t)})$$

where

$$a = \ln((125 - 0.1)/0.1) = 7.13$$

This simple situation, in which there is a linear increase in the environmental resistance with density, that is, logistic growth is valid, seems to hold sufficiently well only for organisms that have a very simple life history. A comparison of the exponential and logistic growth is illustrated graphically in Fig. 4.2.

Figure 4.3 shows the growth rate dN/dt versus the biomass or the number for the logistic growth equation. It has shown a parabolic shape. It is in accordance with the s-shape of the logistic growth equation. The slope has a maximum at an intermediate value of N, but is zero for N = 0 and for N = K.

It is also possible to include harvest (H), which has interest, for instance, in fishery and forest models. The following expression is used:

$$dN/dt = rN\,(1 - N/K) - H$$

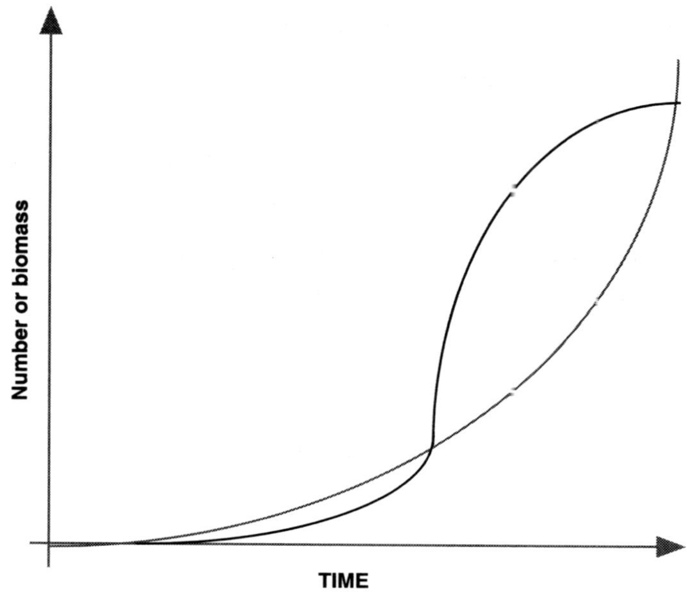

Figure 4.2: Comparison of exponential growth with its linearly increasing slope with logistic growth with its characteristic s-shape.

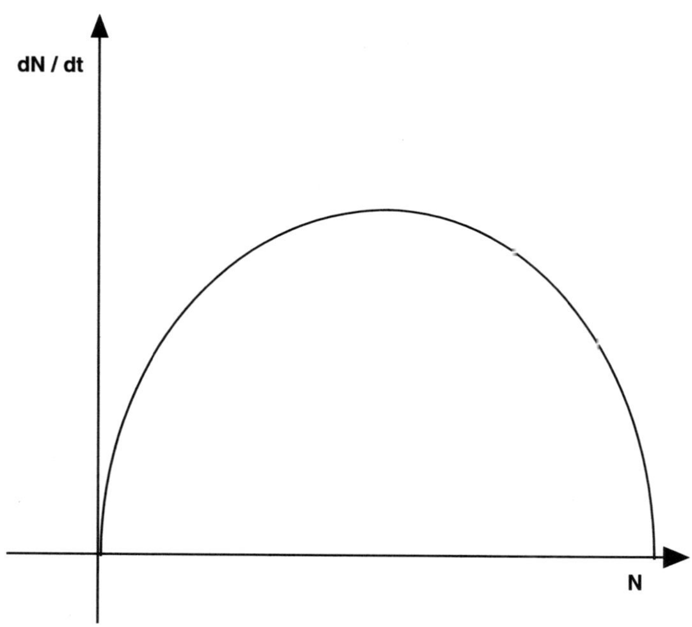

Figure 4.3: dN/dt is plotted versus N for the logistic growth equation.

The harvest H is proportional to N and to the fish effort E:

$$dN/dt = rN\,(1 - N/K) - fEN \tag{4.6}$$

where f is a proportional constant.

This expression has two equilibriums corresponding to dN/dt = 0, $N_1 = 0$ and $N_2 = K\,(1 - fE/r)$. N_2 can be obtained graphically as shown in Fig. 4.4.

If the specific fishing mortality fE is > r, there is no equilibrium value N_2, but only the equilibrium value $N_1 = 0$. For a sustainable harvest $fE/r < 1.0$. The sustainable yield = fEN_2 can be obtained as function of the fishing effort by use of the above shown expression for N_2:

$$\text{Yield} = fEN_2 = fEK\,(1 - fE/r) \tag{4.7}$$

This graph yield 0 f (E) is shown in Fig. 4.5. The optimal effort is obtained by

$$d\text{Yield}/dE = fK - 2f^2EK/r = 0$$

which is leading to

$E = r/(2f)$ and *the maximum yield* = rK/4.

In populations of higher plants and animals that have more complicated life histories, there is likely to be a delayed response. Wangersky and Cunningham [4] have suggested a modification of the logistic equation to include two kinds of time lag: 1) the time needed for an organism to start increasing when conditions are favorable and 2) the time required for organisms to react to

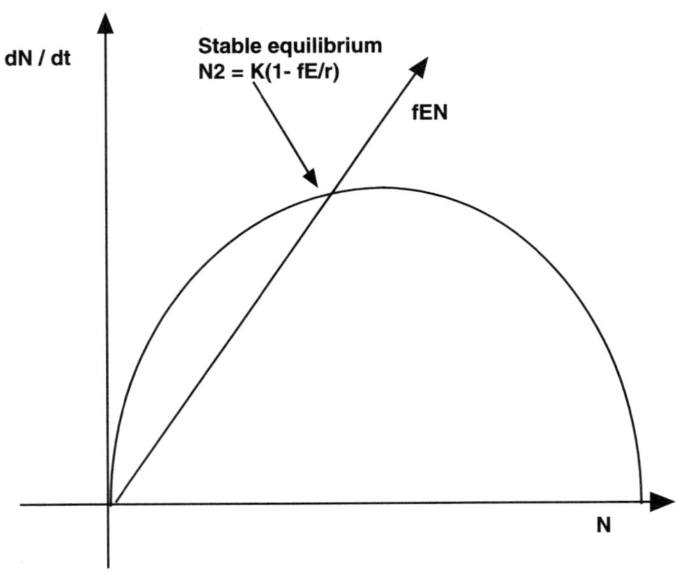

Figure 4.4: The growth rate of the logistic growth as function of N and the fishing mortality as function of N are both plotted. A stable equilibrium is obtained where the two functions are equal. A sustainable fishery will therefore require that the fishing mortality is equal to or less than the increase of N due to the logistic growth.

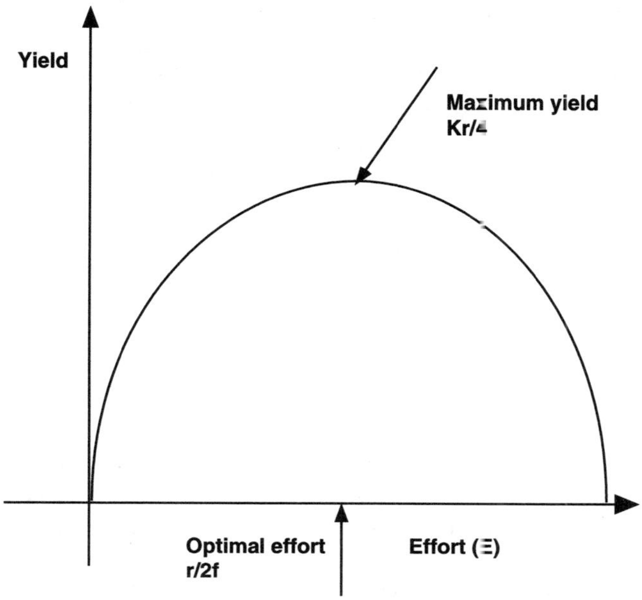

Figure 4.5: Yield is plotted versus fishing effort E. The optimum yield = Kr/4 is obtained by E = r/2f. A sustainable fishery will therefore require that E < r/2f.

unfavorable crowding by altering birth and death rates. If these time lags are $t - t_1$ and $t - t_2$ respectively, we get:

$$dN/dt = r \times N_{t-t1} \times (1 - N_{t-t2}/K) \tag{4.8}$$

Population density tends to fluctuate as a result of seasonal changes in environmental factors or due to factors within the populations themselves (so-called intrinsic factors). We shall not go into details here but just mention that the growth coefficient is often temperature dependent and since temperature shows seasonal fluctuations, it is possible to explain at least some seasonal population fluctuations in density as the temperature changes.

4.4 Interaction between populations

The growth models presented in Section 4.3 presume a constant influence from other populations and that influence is reflected in the selection of parameters. It is, however, unrealistic to assume that interactions between populations are constant. A more realistic model must therefore contain the interacting populations (species) as state variables. For example, in the case of two competing populations we can modify the logistic model and can use the following equations, often termed Lotka–Volterra equation:

$$\frac{dN_1}{dt} = r_1 \cdot N_1 - p_1 N_1 \cdot N_2 \tag{4.9}$$

$$\frac{dN_2}{dt} = p_2 \cdot N_1 \cdot N_2 - d_2 \cdot N_2 \tag{4.10}$$

where N_1 is prey population density, N_2 predator population density, r_1 is the intrinsic (maximal) rate of increase of the prey population (per head), d_2 is the mortality of the predator (per head), and p_1 and p_2 are predation or competition coefficients. p_2 is of course less than p_1 corresponding to an efficiency of the feed resulting from the predation. Each population is limited by the other and in absence of the predator the prey population increases of course exponentially in accordance with a first order reaction. By setting the two right hand sides equal to zero, we find, respectively:

$$N_2 = \frac{r_1}{p_1} \tag{4.11}$$

$$N_1 = \frac{d_2}{p_2} \tag{4.12}$$

Another set of solution is, of course, $N_1 = N_2 = 0$. The two linear eqns (4.11) and (4.12) are plotted in Fig. 4.6 giving dN/dt = 0 isoclines for each species. Below the isoclines, populations will increase and above them they decrease. Therefore, four possible cases result as illustrated in Fig. 4.6.

Thus, each of the two species isocline corresponds to a particular density of the other species. Below the threshold prey density, the predator population will always decrease, whereas above that threshold it will increase. Similarly, the prey population will increase below a particular predator density but decrease above it (see Fig. 4.6). A joint equilibrium exists where the two isoclines cross, but the prey and predator densities do not in general converge to this point, instead any given pair of initial densities results in oscillations of a certain magnitude. The amplitude of fluctuations depends on the initial conditions.

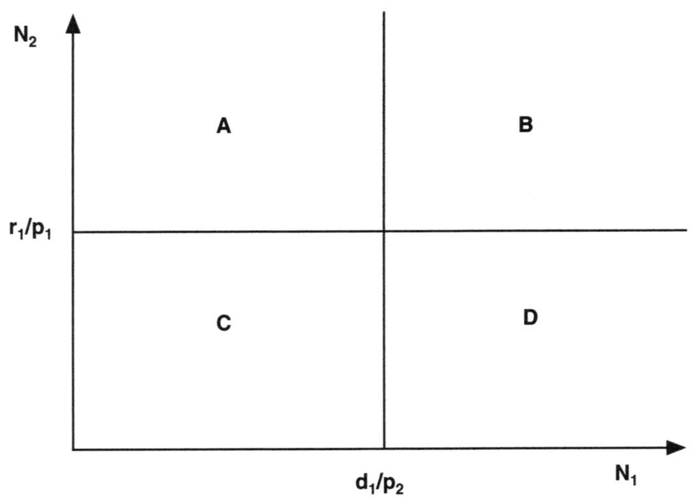

Figure 4.6: Prey–predator isoclines for Lotka–Volterra prey–predator equation. A: both species decrease; B: predator increase, prey decrease; C: prey increase, predator decrease and D: both species increase.

This set of equations is unrealistic since most populations encounter either self-regulations density – dependent feedbacks – or both. The addition of a simple self-damping term to the prey equation results either in a rapid approach to equilibrium or in damped oscillations. Perhaps a more realistic pair of simple equations for modelling the **prey-predator relationship** is:

$$\begin{aligned} \frac{dN_1}{dt} &= r_1 \cdot N_1 - z_1 \cdot N_1^2 - \beta_{12} \cdot N_1 \cdot N_2 \\ \frac{dN_2}{dt} &= \gamma_{21} \cdot N_1 - N_2 - \beta_2 \cdot \frac{N_2^2}{N_1} \end{aligned} \tag{4.13}$$

where r_1, z_1 and so on are coefficients. As can be seen, the prey equation is a logistic expression combined with the effect of the predator, while the predator expression considers a carrying capacity, which is dependent on the prey concentration.

The literature of ecological modelling contains still many papers focusing on the modified Lotka–Volterra equations, but the equations can also be criticized for not following the conservation principle. The increase in the biomass of the predator is less than the decrease in the biomass of the prey. Kooijman [5] has developed many population dynamic models based on the energy conservation principles, which give new and emerging properties of the energy flow in ecosystems. His approach can be recommended when energy is in focus or if a more complex food web is considered.

However, eqn (4.13) can also be easily criticized. The growth term for the predator is seen just as a linear function of the prey concentration of density. Other possible relations are shown in Fig. 4.7. The first relation (a) corresponds to a *Michaelis–Menten expression* (see Chapter 1)

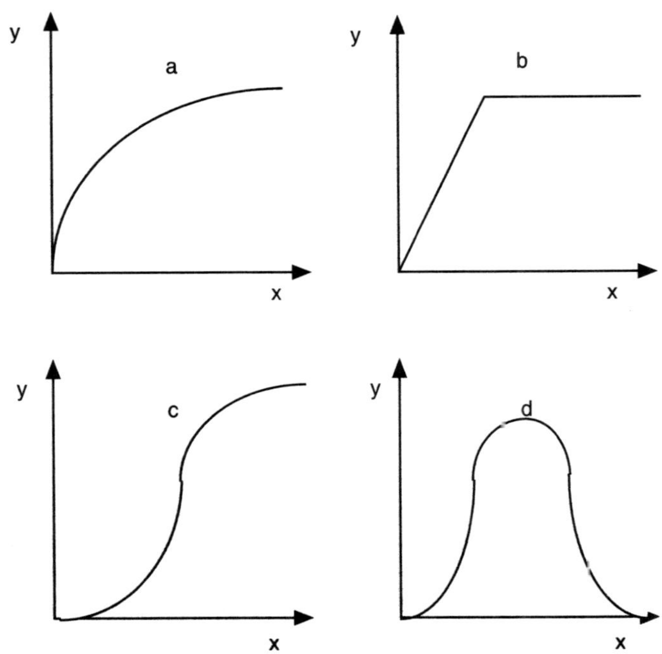

Figure 4.7: Four functional responses [6], where y is number of prey taken per predator per day and x is the prey density.

while the second relation (b) only approximates a *Michaelis–Menten expression* by use of *a first order expression in one interval and a zero order expression in another.* The third relation (c) shown in Fig. 4.7 corresponds to a *logistic expression:* with increasing prey density the predator density first grows exponentially and afterward a damping takes place. This relation has been observed in nature and might be explained as follows: the energy and time used by the predator to capture a prey is decreasing with increasing density of the prey. This implies that the predator not only captures more prey due to increasing density, but also less of the energy consumed is used to capture the next prey. However, at a high prey density another limiting factor starts to be decisive, expressed as a carrying capacity, which is limiting further growth. Behind the carrying capacity is the other limiting factor, for instance, space or nesting sites or predation by a top predator. Therefore, the increase in predator density slows down as it reaches a saturation point at a certain prey density.

The fourth relationship (d) is similar to the often-found relation between growth and pH or temperature. It is characteristic here that the *predator density decreases above a certain prey density.* This response might be explained by the effect of the waste or a toxic compound produced by the prey on the predator. At a certain prey density, the concentration of waste is sufficiently high to have a pronounced negative effect on the predator growth.

Holling [6, 7] has developed many elaborate models of prey–predator relationships. He incorporated time lags and hunger levels to attempt to describe the situation in nature. These models are more realistic, but they are also more complex and require knowledge of more parameters. Besides these complications, we have coevolution of predators and preys.

The prey will develop better and better techniques to escape the predator and the predator will develop better and better techniques to capture the prey. To account for the convolution it is necessary to have a current change of the parameters according to the current selection that takes place.

The effect of **parasitism** is similar to that of predation, but differs from the latter in that members of the host species affected are seldom killed, but may live for sometime after becoming parasitized. This accounts for, by relating the growth and the mortality of the prey (N_1) to the density of the parasites (N_2). The carrying capacity for the parasites is also dependent on the prey density.

The following equations account for these relations and include a carrying capacity of the prey:

$$\begin{aligned} \frac{dN_1}{dt} &= \frac{r_1}{N_2} \cdot N_1 \left(\frac{K_1 - N_1}{K_1} \right) \\ \frac{dN_2}{dt} &= r_2 \cdot N_2 \left(\frac{K_2 \cdot N_1 - N_2}{K_2 \cdot N_1} \right) \end{aligned} \tag{4.14}$$

Symbiotic relationships are modelled with expressions similar to the Lotka–Volterra competition equations simply by changing the signs for the interaction terms.

In nature, interactions among populations often become intricate. The expressions presented above might be of great help in understanding population reactions in nature, but when it comes to the problem of modelling entire ecosystems, they are in most cases insufficient. **Investigations of stability criteria for Lotka–Volterra equations are an interesting mathematical exercise, but can hardly be used to understand the stability properties of real ecosystems or even of populations in nature.**

The experience from the investigations of population stability in nature shows that it is necessary to account for many interactions with the environment to explain observations in real systems, see for example, Deaton and Winebrake [8].

If there is a direct competition between two populations (N_1 and N_2) the following differential equations for the population density (number or biomass per unit of area) may be valid:

$$dN_1/dt = r_1N_1 (1 - N_1/K_1) - m_1N_1 \tag{4.15}$$

and

$$dN_2/dt = r_2N_2 (1 - N_2/K_2) - m_2N_2 \tag{4.16}$$

where r symbolizes the specific growth rate, K the carrying capacity, which is considered different for the two competing populations and m covers the specific mortality rate. If the population density is far from the carrying capacity, which may even be the same for the two populations, we have the following simple expressions, which are often valid when two populations are closely competing about the same natural resources:

$$dN_1/dt = r_1N_1 - m_1N_1 \tag{4.17}$$

and

$$dN_2/dt = r_2N_2 - m_2N_2 \tag{4.18}$$

Adaptation is often a question about a better adjustment to the prevailing conditions for survival. It means that the two specific mortality rates are different, while the specific growth rate may be the same (r). By integration of the two equations and determination of the ratio between the two equations we obtain:

$$N_1(t)/N_2(t) = A_o\, e^{(m2 - m1)\, t} = A_o\, e^{\varepsilon t} \tag{4.19}$$

where ε is the difference in specific mortality.

Illustration 4.2

Find the difference in mortality (survival ability) between *homo sapiens* and *homo neanthertalis* when it is known that the ratio increases from 1 to 10,000 during a period of 10,000 years – from about 40,000 years ago to about 30,000 years ago.

Solution
From eqn (4.19) the following is obtained:

$$10{,}000 = 1 \times e^{\varepsilon 10{,}000}$$

$$\varepsilon 10{,}000 = \ln 10000 = 9.21$$

$$\varepsilon = 0.0009 \text{ or } 0.09\ \%$$

Out of 10,000, 9 more survived among the *homo sapiens* per year.

The stability concept was widely discussed during the 1970s, but today almost all ecologists agree that the stability of an ecosystem is a very complex problem, which cannot be solved by simple methods and at least not by examinations of the stability of only two or three coupled differential equations. It is also acknowledged today that there is no simple relationship between stability and diversity [9–11]. Stability must be considered as a multidimensional concept, because the stability is strongly dependent on which changes that we are concerned with. Some changes – an ecosystem can easily adsorb, some other changes can cause very drastic changes in the ecosystem even by minor changes in the forcing function. The concept ecological buffer capacity is a relevant concept to use in this context, because it is multidimensional. The ecological buffer capacity (B) is defined as:

$$B = \Delta \text{Forcing Function}/\Delta \text{State variable}.$$

The idea behind this concept is the following: if B is high, a significant change in the considered forcing function is needed to provoke a similar change of the considered state variable. It means that the ecosystem has a good ability to adsorb the changes of forcing functions without drastic changes in the ecosystem. If B is low, a minor change in the considered forcing function is sufficient to provoke a significant change of the considered state variable. Notice that a buffer capacity can be determined for each combination of forcing functions and state variables, which makes the concept multidimensional. There is a buffer capacity for each combination of state variable and forcing function.

Figure 4.8 shows a typical relationship between a forcing function and a state variable. Knowledge about such relationships is very important in environmental management. If B is high, it is possible to allow significant changes of the forcing function. If B is low, it is very important to allow only small changes of the forcing function, if the changes of the state variable cannot be accepted, that is, they are leading to a clear decline of the ecosystem conditions from the ecosystem management point of view. If the corresponding changes of the state variable are acceptable, that is, they are not leading to the decline of the ecosystem conditions, the minor changes of the forcing function is of course acceptable.

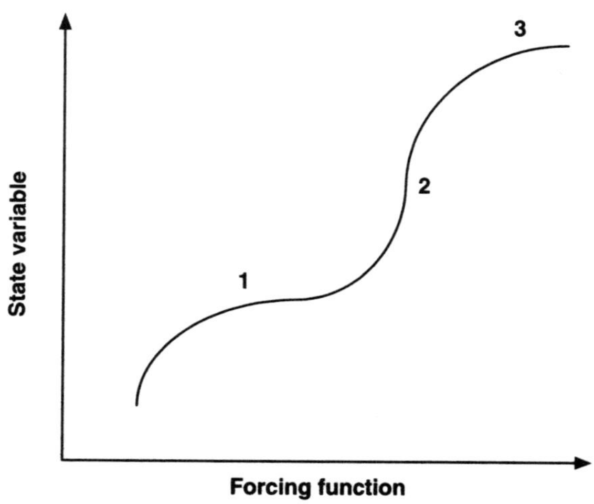

Figure 4.8: A state variable is plotted versus a forcing function. At 1 and 3, the buffer capacity is high and at 2, the buffer capacity is low. It is important in environmental management to assess when the buffer capacity is low and high to be able to avoid drastic changes of ecosystems.

4.5 Matrix models

Another important aspect of modelling population dynamics is the influence of the age distribution, which shows the proportion of the population belonging to each age class. If a population has unchanged growth rate, life expectation and mortality, it will eventually reach a stable age distribution; it means that the percentage of organisms in each age class remains the same. Recruitment into every age class is exactly balanced by its loss due to mortality and aging.

The growth equations presented in Section 4.3 and all the equations assume that the population has a stable age distribution. The intrinsic rate of increase or growth (r), the generation time (T) and the reproductive value are conceptually independent of the age distribution, but might, of course, be different for population of the same species with different age distributions. Therefore, the models presented in the previous two sections did not need to consider age distribution, although the parameters in actual cases, of course, should reflect the actual age distribution.

A model predicting the future age distribution was developed by Leslie [12]. The population is divided into n + 1 equal age groups – group 0, 1, 2, 3,….n. The model is presented by the following matrix equation:

$$\begin{matrix} f_0 & f_1 & f_2 & \cdots & f_{n-1} & f_n \\ p_0 & 0 & 0 & \cdots & 0 & 0 \\ 0 & p_1 & \cdots & \cdots & 0 & 0 \\ \cdots & \cdots & \cdots & \cdots & \cdots & \cdots \\ \cdots & \cdots & \cdots & \cdots & \cdots & \cdots \\ 0 & 0 & 0 & \cdots & p_{n-1} & 0 \end{matrix} \quad \begin{matrix} n_{t,0} \\ n_{t,1} \\ n_{t,2} \\ . \\ . \\ n_{t,n} \end{matrix} \quad \begin{matrix} n_{t+1,0} \\ n_{t+1,1} \\ n_{t+1,2} \\ . \\ . \\ n_{t-1,n} \end{matrix}$$

The number of organisms in the various age classes at time t + 1 is obtained by multiplying the number of animals in the age classes at time t by a matrix, which expresses the fecundity and survival rates for each age class. f_0, f_1, f_2, …..f_n give the reproduction in the age group number i and P_0, P_1, P_2, P_3, P_4, ……P_n represent the probability that an organism in the i'th age group still will be alive after promotion to the (i + 1)th group.

The model can be written in the following form:

$$\boldsymbol{A} \times \boldsymbol{a}_t = \boldsymbol{a}_{t+1} \tag{4.20}$$

where $\boldsymbol{A}$ is the matrix, $\boldsymbol{a}_t$ is the column vector representing the population age structure at time t and $\boldsymbol{a}_{t+1}$ is a column vector representing the age structure at time t + 1. This equation can be extended to predict the age distribution after $\boldsymbol{k}$ periods of time:

$$\boldsymbol{a}_{t+k} = \boldsymbol{A}^k \times \boldsymbol{a}_t \tag{4.21}$$

The matrix $\boldsymbol{A}$ has n + 1 possible eigenvalues and eigenvectors. Both the largest eigenvalues (λ) and the corresponding eigenvectors are ecologically meaningful λ gives the rate at which the population size is increased:

$$\boldsymbol{A} \times \boldsymbol{v} = \lambda \times \boldsymbol{v} \tag{4.22}$$

where $\boldsymbol{v}$ is the stable age structure and λ is the intrinsic rate of natural increase. The corresponding eigenvector indicates as seen the stable structure of the population.

Illustration 4.3

Usher [13] has given a very illustrative example on the use of matrix models. The model is based on data provided for the blue whale before sharp changes in survival rates were initiated.

The eigenvalue can be used to find the number of individuals that can be removed from a population to maintain the same number in each age class. It can be shown that the following equation is valid:

$$H = 100\ (\lambda - 1)\lambda$$

where H is the percentage of the population, which can be removed for instance by hunting.

The blue whales reach maturity between 4 and 7 years of age. They have a gestation period of about 1 year. A single calf is born and is nursed for about 7 months. In average, not more than one calf is born to a female every 2 years. The numbers of the two sexes are approximately equal. Survival rates are about 0.7 for every 2 years for the first 10 years and 0.78 for whales above 12 years. We divide the population into 7 groups with a two-year-period for the first six groups and the age of 12 years and above as the seventh group. The fecundity for the first two groups is zero according to the obtained information. The third group has a fecundity of 0.19 and the fourth group of 0.44. The maximum fecundity of 0.50 is reached at the age of 8–11 years. The fecundity of the last group is 0.45.

Find the intrinsic rate of natural increase, the stable structure of the whale population and the harvest, which can be taken to maintain a stable population size.

Solution

The eigenvalue can be found either by an iterative method or by plotting the number of whales (totally or for each age class separately) versus the time-period. The slope of this plot will after a stabilization period correspond to r, the intrinsic rate of increase or ins. We find by these methods that r = 0.0036/year or λ = antilog 0.0036 (for 1 year) or 1.00362 × 2 = 1.0072 for 2 years. The corresponding eigenvector is found by use of eqn (4.22) to be: a = [1000, 764, 584, 447, 341, 261, 885] as the Leslie matrix is:

$$\begin{matrix} 0 & 0 & 0.19 & 0.44 & 0.50 & 0.50 & 0.45 \\ 0.77 & 0 & 0 & 0 & 0 & 0 & 0 \\ 0 & 0.77 & 0 & 0 & 0 & 0 & 0 \\ 0 & 0 & 0.77 & 0 & 0 & 0 & 0 \\ 0 & 0 & 0 & 0 & 0.77 & 0 & 0 \\ 0 & 0 & 0 & 0 & 0 & 0.77 & 0 \\ 0 & 0 & 0 & 0 & 0 & 0.77 & 0.78 \end{matrix}$$

The harvest that can be taken from the population is estimated to be:

$$H = 100\ (\lambda - 1)/\lambda = 0.71\%$$

every 2 years or about 0.355% every year.

If the harvest exceeds this value, the population will decline. Population models of r-strategies might generally cause some more difficulties to develop than models of K-strategies due to the high sensitivity of the fecundity. The number of offsprings might be known quite well, but the number of survivors to be included in the first age class and the number of recruits is difficult to predict. This is the central problem of fish population dynamics, since it represents nature's regulation of population size.

4.6 Metapopulation models

A regional set of local populations, which are occupying isolated habitat patches but are interconnected by dispersal movements, are denoted as metapopulations; see an example in Fig. 4.9. All the local populations have a finite possibility of becoming extinct. Even if the local population is fairly large extinction may still occur, for instance, through catastrophic events. The dispersal movements are essential to re-establish populations that have faded or crashed to zero numbers. Species that are widely distributed in many local populations have reduced the likelihood to become extinct regionally.

An increasing fragmentation of landscapes has taken place due to the human expansion. Populations that were formerly continuously distributed have become broken into separate localized groupings. Dispersal may even be inhibited by hazards in traversing the human transformed areas separating suitable habitats. Metapopulation models assess the risks of species extinctions as a consequence of such fragmentations and identify how actions such as providing dispersal corridors can reduce the risks.

The metapopulation concept was formulated by Levins [14] and further modified by Hanski [15, 16]. P is the proportion of sites occupied by populations, E is the extinction rate of these populations and C is the colonization rate of vacant sites by migrants from occupied patches. The change over time in the proportion of patches (dP/dt), occupied is a matter of balance between the colonization and the extinction:

$$dP/dt = CP\,(1 - P) - EP \tag{4.23}$$

The equilibrium proportion P_{eq} is given by

$$P_{eq} = 1 - E/C \tag{4.24}$$

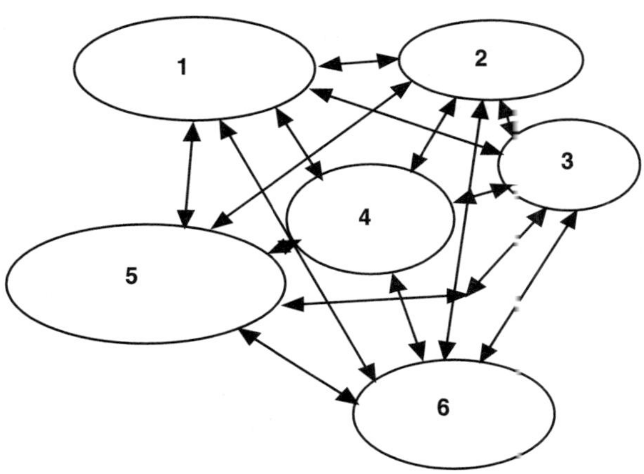

Figure 4.9: Schematic outline of a metapopulation. The populations occupy isolated habitats that are connected by patches.

As seen from eqn (4.24), the patch occupancy will become zero if the extinction rate exceeds the colonization rate.

Figure 4.10 shows a STELLA diagram for metapopulation models according to eqn (4.23) plus an introduction of disturbances as function of time. Figure 4.11 shows a result of a simulation where there are two small disturbances of patches and an interruption of the colonization from day 83 to the day of the colonization due to human activities. The simulation time is 500 days. The equations are given in Table 4.2 in the STELLA format.

4.7 Modelling infectious diseases

Population models of disease dynamics has – as the focal state variable – the proportion of the host population that is infected. A simple infection model is shown in Fig. 4.12. Susceptible hosts (S) become infected at rate b. After a period, the infective hosts either recover maybe with long-lasting immunity or die. The number of infective hosts is thereby reduced at a rate corresponding to m + r, where m is the mortality rate and r is the recovery rate. The disease is spreading because of contact between infected and susceptible hosts. The following equation can be used to express the number of infected hosts, I:

$$dI/dt = bSI - rI - mI \tag{4.25}$$

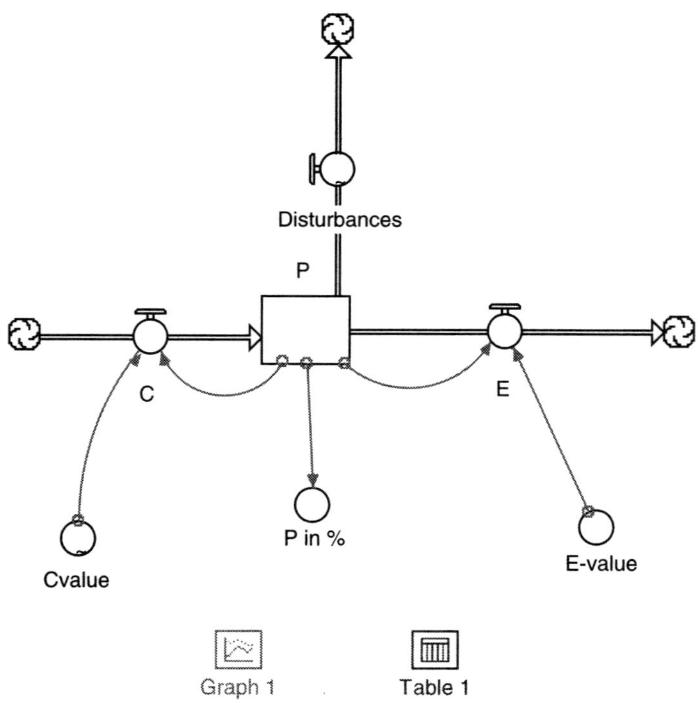

Figure 4.10: A STELLA diagram of a metapopulation model with a colonization rate, that is, 0 between day 83 and day 225 and with two short and minor disturbances (removal of patches). The simulation time is 500 days.

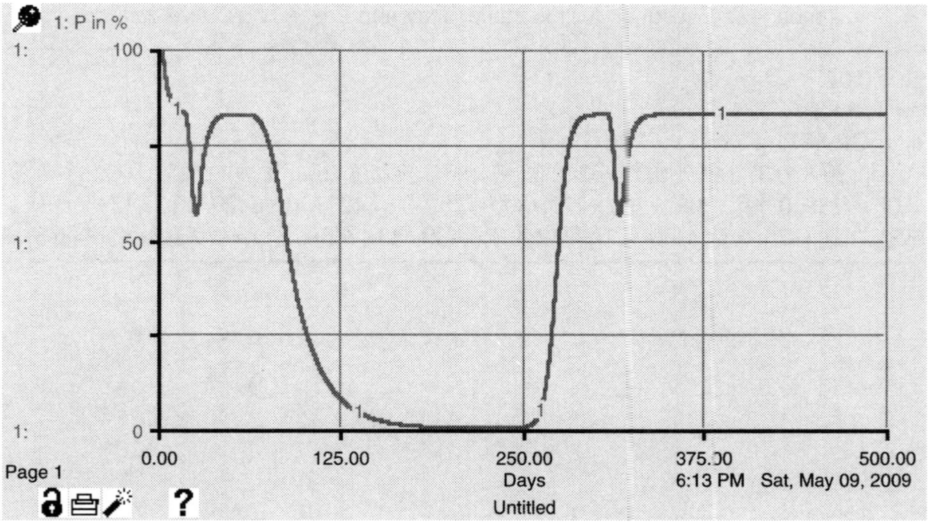

Figure 4.11: The results of the model shown in Fig. 4.10. The colonization is interrupted from day 83 to day 225. The occupied patches are recovering at the equilibrium (see eqn (4.24)). As C is 0.3 (except from day 83 to 225) and E is 0.05, the equilibrium 1 – 1/6 or 83.4%.

Table 4.2: Equations for the model shown in Fig. 4.12.

P (t) = P (t – dt) + (C – E – Disturbances) × dt
INIT P = 1

INFLOWS:
C = C value × P × (1 – P)
OUTFLOWS:
E = E value × P
Disturbances = GRAPH (TIME)
(0.00, 0.00), (5.05, 0.00), (10.1, 0.00), (15.2, 0.00), (20.2, 0.1), (25.3, 0.00), (30.3, 0.00), (35.4, 0.00), (40.4, 0.00), (45.5, 0.00), (50.5, 0.00), (55.6, 0.00), (60.6, 0.00), (65.7, 0.00), (70.7, 0.00), (75.8, 0.00), (80.8, 0.00), (85.9, 0.00), (90.9, 0.00), (96.0, 0.00), (101, 0.00), (106, 0.00), (111, 0.00), (116, 0.00), (121, 0.00), (126, 0.00), (131, 0.00), (136, 0.00), (141, 0.00), (146, 0.00), (152, 0.00), (157, 0.00), (162, 0.00), (167, 0.00), (172, 0.00), (177, 0.00), (182, 0.00), (187, 0.00), (192, 0.00), (197, 0.00), (202, 0.00), (207, 0.00), (212, 0.00), (217, 0.00), (222, 0.00), (227, 0.00), (232, 0.00), (237, 0.00), (242, 0.00), (247, 0.00), (253, 0.00), (258, 0.00), (263, 0.00), (268, 0.00), (273, 0.00), (278, 0.00), (283, 0.00), (288, 0.00), (293, 0.00), (298, 0.00), (303, 0.00), (308, 0.00), (313, 0.1), (318, 0.00), (323, 0.00), (328, 0.00), (333, 0.00), (338, 0.00), (343, 0.00), (348, 0.00), (354, 0.00), (359, 0.00), (364, 0.00), (369, 0.00), (374, 0.00), (379, 0.00), (384, 0.00), (389, 0.00), (394, 0.00), (399, 0.00), (404, 0.00), (409, 0.00), (414, 0.00), (419, 0.00), (424, 0.00), (429, 0.00), (434, 0.00), (439, 0.00), (444, 0.00), (449, 0.00), (455, 0.00), (460, 0.00), (465, 0.00), (470, 0.00), (475, 0.00), (480, 0.00), (485, 0.00), (490, 0.00), (495, 0.00), (500, 0.00).

Table 4.2: Equations for the model shown in Fig. 4.12. *(Continued)*

E value = 0.05
Pin % = 100 × P
C value = GRAPH (TIME)
(0.00, 0.3), (20.8, 0.3), (41.7, 0.3), (62.5, 0.29), (83.3, 0.00), (104, 0.00), (125, 0.00), (146, 0.00), (167, 0.00), (188, 0.00), (208, 0.00), (229, 0.00),(250, 0.3), (271, 0.3), (292, 0.3), (313, 0.3), (333, 0.3), (354, 0.3), (375, 0.3), (396, 0.3), (417, 0.3), (437, 0.3), (458, 0.3), (479, 0.3), (500, 0.3).

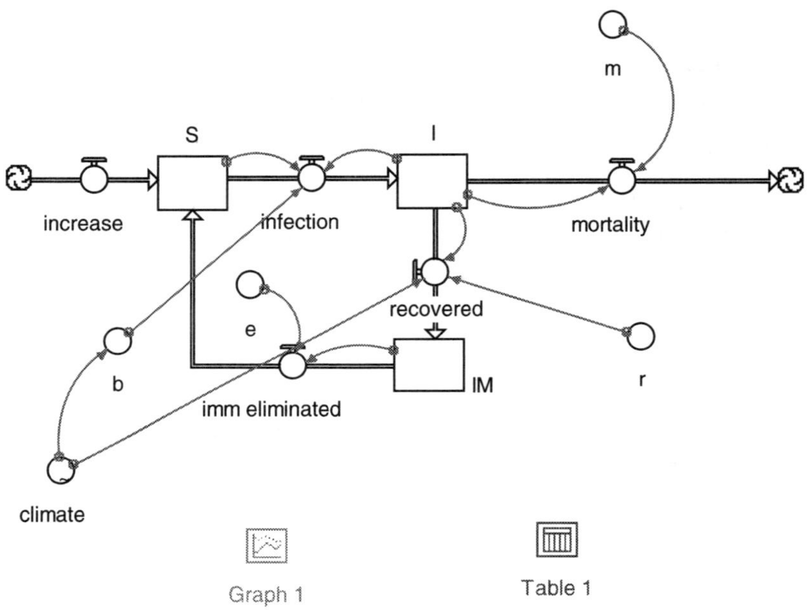

Figure 4.12: A conceptual STELLA diagram for an infection model is shown. The model has three state variables: the number of susceptible hosts, the number of infected host and the number of immune hosts. The infection rate is b, the mortality rate is m, the recovery rate is r and the rate of immunity elimination is e.

The duration of recovery (D) is the inverse of the rate of recovery from the infection:

$$D = bN/r \tag{4.26}$$

where N is the susceptible hosts, which initially may be equal to the total population, denoted N.

Notice that eqn (4.25) implies that the infection will spread very fast, because it is at a rate that is S times I. Figure 4.12 shows a conceptual diagram of an infection model with three state variables: susceptible hosts, infected hosts and immune hosts. The spreading of the infection follows eqn (4.25). The infected hosts either die at a rate m or recover at a rate r. The immunity is eliminated at a rate e, which implies that immune hosts are transferred to the susceptible hosts at a rate e.

Table 4.3: Equations for the model shown in Fig. 4.12.

I(t) = I (t – dt) + (infection – mortality – recovered) × dt
INIT I = 1

INFLOWS:
infection = b × S × I
OUTFLOWS:
mortality = m × I
recovered = I × r × climate
IM(t) = IM (t – dt) + (recovered – imm eliminated) × dt
INIT IM = 0

INFLOWS:
recovered = I × r × climate
OUTFLOWS:
imm eliminated = IM × e
S(t) = S (t – dt) + (imm eliminated + increase – infection) × dt
INIT S = 1000000

INFLOWS:
imm eliminated = IM × e
increase = 200
OUTFLOWS:
infection = b × S × I
b = 0.000001 × climate
e = 0.025
m = 0.002
r = 0.05
climate = GRAPH (TIME)
(0.00, 1.40), (20.4, 1.91), (40.8, 2.00), (61.2, 1.80), (81.6, 1.35), (102, 0.85), (122, 0.61), (143, 0.4), (163, 0.32), (184, 0.29), (204, 0.36), (224, 0.43), (245, 0.81), (265, 1.08), (286, 1.26), (306, 1.46), (327, 1.60), (347, 1.75), (367, 1.86), (388, 1.96), (408, 2.00), (429, 1.87), (449, 1.53), (469, 0.86), (490, 0.62), (510, 0.44), (531, 0.35), (551, 0.3), (571, 0.34), (592, 0.5), (612, 0.73), (633, 0.97), (653, 1.39), (673, 1.75), (694, 1.94), (714, 2.00), (735, 1.97), (755, 1.86), (776, 1.68), (796, 1.43), (816, 1.18), (837, 0.9), (857, 0.6), (878, 0.42), (898, 0.3), (918, 0.35), (939, 0.55), (959, 1.01), (980, 1.28), (1000, 1.45).

The influence of the climate is included in the model. The auxiliary variable "climate" (see Fig. 4.12) follows a sin like curve with higher values during winter (maximum in February with an approximate value of 2.0) and lower during summer with an approximate value of 0.3. The equations of the model are shown in Table 4.3. The result of a simulation with the duration of 1000 days is shown in Fig. 4.13. As expected, the number of infected hosts are increasing very rapidly, although b is only 0.000001. The number of susceptible hosts is 1,000,000 at time = 0 and the number of infected hosts has as initial value 1.0. These numbers could be realistic for an epidemic influenza. Notice that the number of infected hosts after the peak has been reached is decreasing, but with fluctuations according to the auxiliary variable "climate". The fluctuations of the number of immune hosts and infected hosts are opposite, with a maximum for infected hosts when the number of immune hosts are in minimum.

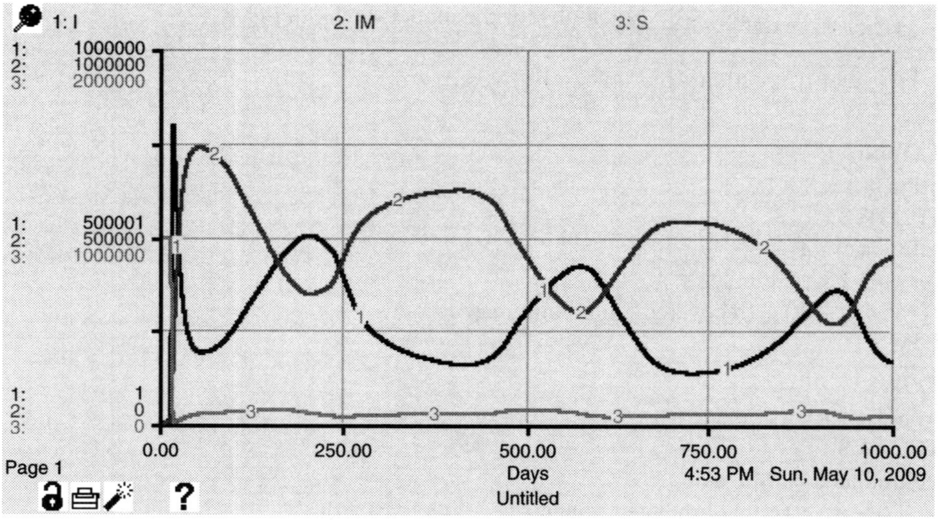

Figure 4.13: The simulation results of the model shown in Fig. 4.12. The equation is applied in Table 4.3. Notice the rapid increase due to the equation dI/dt = bIS. The peak of infections is after 10 days. The fluctuations of all three state variables, particularly for I and IM, are due to the auxiliary variables "climate".

References

[1] Lotka, A.J., *Elements of Mathematical Biology*, Dover: New York, 1956.

[2] Volterra, V., Auctuations in the abundance of a species considered mathematically. *Nature*, **188**, pp. 558–560, 1926.

[3] Jørgensen, S.E., Chon, T. & Recknagel, F., *Handbook of Ecological Modelling and Informatics*, WIT Press: Southampton, 2009.

[4] Wangersky, P.J. & Cunningham, W.J., Time lag in population models. *Cold Spring Harbor Symp. Quant. Biol.*, **42**, pp. 329–338, 1957.

[5] Kooijman, S.A.L.M., *Dynamic Energy and Mass Budgets in Biological Systems*, Cambridge University Press: Cambridge, 2000.

[6] Holling, C.S., Some characteristics of simple types of predation and parasitism. *Canad. Entomol.*, **91**, pp. 385–398, 1959.

[7] Holling, C.S., The functional response of invertebrate predators to prey density. *Mem. Entomol. Soc. Canada*, **48**, pp. 1–87, 1966.

[8] Deaton, M.L. & Winebrake, J.J., *Dynamic Modeling of Environmental Systems*, Springer Verlag: Berlin, 2000.

[9] May, R.M., Ecosystem patterns in randomly fluctuating environments. *Progr. Theor. Biol.*, **3**, pp. 1–50, 1974.

[10] May, R.M., Patterns of species abundance and diversity (Chapter 4). *Ecology and Evolution of Communities*, eds. M.L. Cody & J.M. Diamond, Harvard University Press: Cambridge, MA, pp. 81–120, 1975.

[11] May, R.M., *Stability and Complexity in Model Ecosystems*, 3rd edn, Princeton University Press: Princeton, NJ, 1977.

[12] Leslie, P.H., On the use of matrices in certain population mathematics. *Biometrika*, **33**, pp. 183–212, 1945.

[13] Usher, M.B., Developments in the Leslie matrix model. *Mathematical Models in Ecology*, ed. J.N.R. Jeffers, Blackwells: Oxford, pp. 29–60, 1972.

[14] Levins, R., Some demographic and genetic consequences of environmental heterogeneity for biological control. *Bulletin of the Entomological Society of America*, **15**, pp. 237–240, 1969.

[15] Hanski, I., Metapopulation dynamics in fragmented landscapes. *Trends in Ecology and Evolution*, **9**, pp. 131–195, 1994.

[16] Hanski, I., *Metapopulation in Ecology*, Oxford University Press: Cambridge, 1999.

Exercises/problems – Chapter 4

1. Set up a STELLA model representing Lotka–Volterra equations. How is it possible to consider the conservation principles, which is a prerequisite for the application of STELLA?
2. Express the model in Illustration 4.1 by STELLA.
3. Give an overview of factors that may be able to limit the carrying capacity of a population.
4. Make a conceptual diagram of a four species model based on population interactions representing prey–predator top predators and two competing top predators.
5. Mention at least three reasons why the Lotka–Volterra model could be an unrealistic description of the population dynamic.
6. A fish culture has a carrying capacity of 80 g/l. Set up a logistic growth equation for the fish culture, when the initial concentration is 1 g/l at day 0 and the concentration 2 g/l is obtained after 12 days. How long does it take to increase the concentration from 20 g/l to 40 g/l? Find an equation that expresses the doubling time as function of the time.
7. Set up a matrix model for a bird population that has the following characteristics:
   (a) Life span of 7 years;
   (b) Four eggs from the second year per pair, increasing to five eggs the third year and six eggs the following years;
   (c) The mortality is 30% the first year, 20% the following years, except the last year where it is 100%.

   What is the steady state age distribution?
8. Mention at least three concrete cases, where metapopulation models could be used to solve environmental management problems.
9. Reconstruct the model in Fig. 4.12 and use the model to assess the role (i.e., the influence on the results) of the following important parameters:
   (a) mortality (m),
   (b) rate of recovery (r),
   (c) rate of immunity elimination (e),
   (d) infection rate (b).

   Include a time lag in the infection process. How does the length of the time lag influence the results?
10. The following equation is valid for a fish population. dN/dt = 0.03 × number of fish × (1 – fish/10^7). The fishing effort (E) is 0.2 and f = 0.7. Find by use of a graph dN/dt for the stable equilibrium. What is the maximum yield and what is the optimal effort?
11. Let us consider the eutrophication according to the model presented in Section 3.4.
    Give a qualitative estimation of the influence on the buffer capacity change in phosphorus loading/phytoplankton concentration of the following factors (forcing functions):
    (a) temperature,
    (b) nitrogen loading,
    (c) oxygen concentration at the sediment surface (reducing the release of phosphorus to the water phase,
    (d) fishery of top carnivorous species,
    (e) water retention time.

    Set up relevant buffer capacities using the forcing functions a–e and phytoplankton as state variable in the buffer capacity expression.

CHAPTER 5

Ecotoxicological models

5.1 Introduction

Ecotoxicological models are widely used in environmental risk assessment (ERA). In accordance to the suggested procedure for ERA, the distribution of a toxic substance in the environment is found by the development of a model. The model gives us a result: the concentrations in different compartments: soil, plants, birds, fish, groundwater etc. The concentration in a compartment found by the model is called predicted environmental concentration (PEC). Parallel to the model development, the highest concentration that is not causing an unacceptable negative effect is determined. This concentration is divided by an assessment factor or uncertainty factor to account for the uncertainty. The result is a concentration called predicted no effect concentration (PNEC) and it is found for various different environmental compartments such as water, soil, air, biota etc. The ratio between PEC and PNEC determines the risk. If the ratio is more than 1.0 there is a significant risk and thereforethe release of the toxic substance to the environment must be reduced. If on the other hand the ratio is less than 0.1, the risk is negligible. For ratios between 0.1 and 1.0, it is recommendable to have an open eye on the development of the ratio to be able, in due course of time, to stop or reduce the risk. The usually applied procedure for ERA is shown in Fig. 5.1.

This chapter on ecotoxicological models is, however, not only included in this book on modelling due to the wide use in ERA, but also because ecotoxicological models illustrate very clearly the general advantages of model applications on the one side and demonstrate the uncertainty of the model results on the other side. It is very important as already emphasized in Chapter 1 to assess and accept the uncertainty of models. Models are powerful tools in environmental management and in ecological research, but the benefits gained by use of models are lost if the users are not accepting, considering and preferably also determining the uncertainty. The uncertainty is best determined by a validation of the model as discussed in Chapter 1 and clearly illustrated for a eutrophication model in Chapter 3. Ecotoxicological models have usually a high uncertainty, because they are usually simple and not accounting for all the many not completely negligible details. They have also parameter values with a high uncertainty. It is, however, by the application of the assessment factor which is possible to take the uncertainty into account in ERA and other similar applications.

Toxic substance and ecotoxicological models are most often but not necessarily always biogeochemical models, because they attempt to describe the mass flows of toxic substances.

Toxic substance models differ from other ecological models in that:

1. The need for parameters to cover all possible toxic substance models is huge and general estimation methods are therefore widely used.

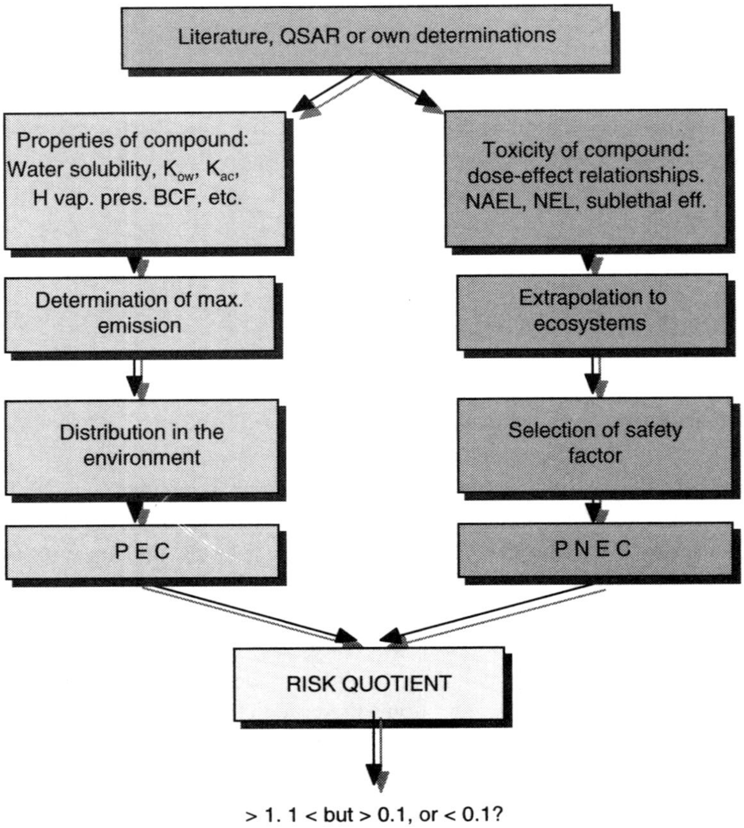

Figure 5.1: The usually applied procedure for ERA is shown. It is necessary to determine the distribution of the toxic substance in the environment by development of an environmental–ecological model. Reproduced from [2].

2. The safety margin, assessment factors, should be high, for instance, expressed as the ratio between the predicted concentration and the concentration that gives undesired effects. The selection of the assessment factor is mainly a question of our knowledge about the effect of the chemical and the uncertainty of the parameter values.
3. They require possible inclusion of an effect component, which relates the output concentration to its effect. It is easy to include an effect component in the model; it is, however, often a problem to find a well-examined relationship to base it on.
4. There is a need for simple models due to points 1 and 2, and due to our limited knowledge of process details, parameters, sublethal effects, antagonistic and synergistic effects.

It may be an advantage to emphasize the use of the generally accepted modelling approach for toxic substance models according to the procedure presented in Chapter 1, Fig. 1.3:

1. Obtain the best possible knowledge about the possible processes of the considered toxic substances in the ecosystem. As far as possible, the knowledge about the quantitative role of the processes should be obtained.

2. Attempt to get as many parameter values as possible from the literature and/or from our own experiments (*in situ* or in the laboratory).
3. Estimate all parameters by the methods presented in Section 5.3.
4. Compare the results from the above points 2 and 3, and attempt to explain discrepancies. This comparison will be able to give a first estimation of the uncertainty associated with the parameters, which is often the crucial uncertainty.
5. Estimate which processes and state variables would be feasible and relevant to include in the model. If there is the slightest doubt then include at this stage rather too many processes and state variables than too few.
6. Use a sensitivity analysis to evaluate the significance of the individual processes and parameters. This may often lead to further simplification.

To summarize, ecotoxicological models differ from ecological models in general by [1]:

1. being most often more simple,
2. requiring more parameters,
3. a wide use of parameter estimation methods,
4. a possible inclusion of an effect component.

Ecotoxicological models may be divided into five classes according to their structure. The five classes also illustrate the possibilities of simplification which is urgently needed for ecotoxicological models – as already touched on a few times.

5.2 Five classes of ecotoxicological models

5.2.1 Food chain or food web dynamic models

This class of models considers the flow of toxic substances through the food chain or food web. Such models will be relatively complex and contain many state variables. The models will also contain many parameters. This type of model will typically be used when many organisms are affected by the toxic substance or the entire structure of the ecosystem is threatened by the presence of a toxic substance. Because of the complexity of these models, they have not been used widely. They are similar to the usually more complex eutrophication models (see Chapter 3), which consider the flow of nutrients through the food chain or even through the food web. Sometimes they are even constructed as submodels of a eutrophication model. A simplification is hardly possible for this type of model, because it is the aim of the model to describe and quantify the bioaccumulation through the entire food chain step by step. The result of an aquatic model could, for instance, be the biomagnification from, for instance, the water to the fish via phytoplankton, zooplankton and planktivorous fish. Biomagnification is equal to the ratio of concentration in the carnivorous fish/the concentration in water.

5.2.2 Static models of the mass flows of toxic substances

If the seasonal changes are minor, or of minor importance, a static model of the mass flows will often be sufficient to describe the situation and even to show the expected changes if the input of toxic substances is reduced or increased. An ecotoxicological static model has, of course, the general advantages and disadvantages of static models: they require less data and are simple to develop; but they cannot, of course, reveal the dynamic of the system.

This type of model is based upon a mass balance as clearly seen from the example in Fig. 5.2 (reproduced from [2]). It will often, but not necessarily, contain more trophic levels, but the modeller is frequently concerned with the flow of the toxic substance through the food chain. The example in Fig. 5.2 considers only one trophic level. If there are some seasonal changes, this type, which usually is simpler than type one, can still be an advantage to use, for instance, if the modeller is concerned with the worst case or the average case and not with the changes.

5.2.3 A dynamic model of a toxic substance in one trophic level

It is often only the toxic substance concentration in one trophic level that is of concern. This includes the zero trophic level, which is understood as the medium – either soil, water or air. Figure 5.3 (reproduced from [2]) gives an example. Here the main concern is the dichloro-diphenyl-trichlorethylene (DDT) concentration in fish, where there may be such high concentration of DDT that, according to the World Health Organization standards, they are not recommended for human consumption. The model can, therefore, be simplified by not including the entire food chain but only the fish. The biomagnification through the food chain is taken into consideration by use of a magnification factor, which may be found in the literature or estimated from the knowledge about the food chain. Some physical–chemical reactions in the water phase are still of importance and they are included as shown on the conceptual diagram Fig. 5.3. As seen from these examples, simplifications are often feasible when the problem is well defined; including which component is most sensitive to toxic matter and which processes are most important for concentration changes.

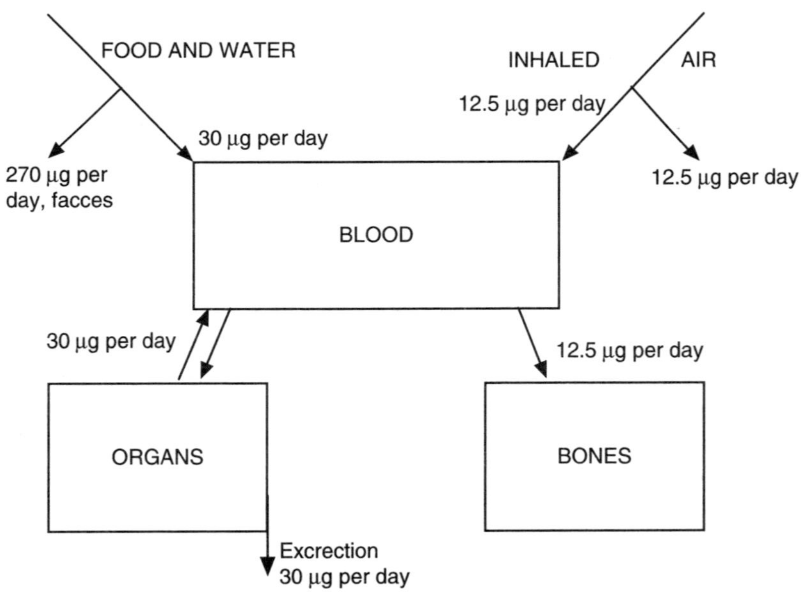

Figure 5.2: A static model of the lead uptake by an average Dane in the 1970s before the use of lead in gasoline was banned.

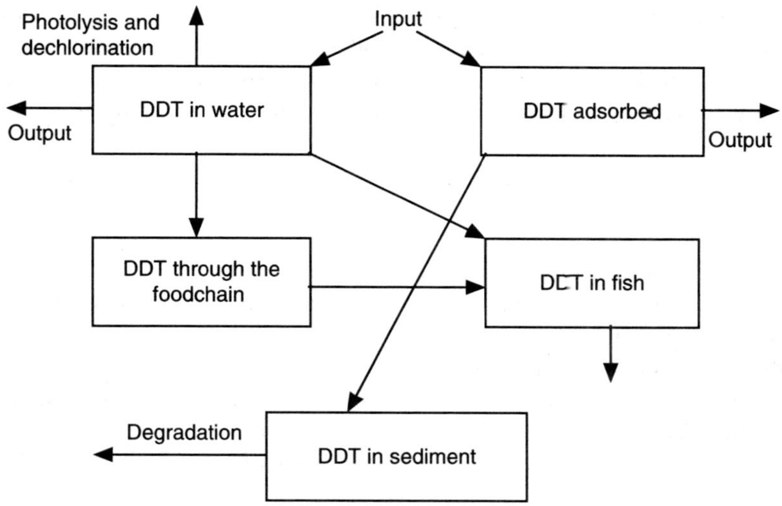

Figure 5.3: Conceptual diagram of a simple DDT model.

5.2.4 Ecotoxicological models in population dynamics

Population models are biodemographic models and have, therefore, numbers of individuals or species as state variables. The simple population models consider only one population. The growth of the population is a result of the difference between natality and mortality (see Chapter 4):

$$dN/dt = B \times N - M \times N = r \times N, \tag{5.1}$$

where N is the number of individuals; B is the natality, that is, the number of new individuals per unit time and per unit population; M is the mortality, that is, the number of organisms that die per unit time and per unit population and r is the increase in the number of organisms per unit time and per unit population, and is equal to B – M. B, N and r are not necessarily constants as in the exponential growth equation, but may be dependent on N, the carrying capacity and other factors.

The concentration of a toxic substance in the environment or in the organisms may influence the natality and the mortality, and if the relation between a toxic substance concentration (tx) and these population dynamic parameters is included in the model, it becomes an ecotoxicological model of population dynamics:

$$dN/dt = B\,(tx) \times N - M\,(tx) \times N = r\,(tx) \times N. \tag{5.2}$$

Population dynamic models may include two or more trophic levels and ecotoxicological models will include the influence of the toxic substance concentration on natality, mortality and interactions between these populations. In other words, an ecotoxicological model of population dynamics is a general model of population dynamics with the inclusion of relations between toxic substance concentrations and some important model parameters.

5.2.5 Ecotoxicological models with effect components

Although class 4 models already may include relations between concentrations of toxic substances and their effects, these are limited to, for instance, population dynamic parameters, not to a final assessment of the overall effect. In comparison, class 5 models include more comprehensive relations between toxic substance concentrations and their effects. These models may include not only lethal and/or sublethal effects but also effects on biochemical reactions or on the enzyme system. The effects may be considered on various levels of the biological hierarchy from the cells to the ecosystems.

In many problems, it may be necessary to go into more detail on the effect to answer the following relevant questions:

1. Does the toxic substance accumulate in the organism?
2. What will be the long-term concentration in the organism when uptake rate, excretion rate and biochemical decomposition rate are considered?
3. What is the chronic effect of this concentration?
4. Does the toxic substance accumulate in one or more organs? At which concentation?
5. What is the transfer between various parts of the organism?
6. Will decomposition products eventually cause additional effects?

A detailed answer to these questions may require a model of the processes that take place in the organism and a translation of concentrations in various parts of the organism into effects. This implies, of course, that the intake = (uptake by the organism) × (efficiency of uptake) is known. Intake may either be from water or air, which also may be expressed (at steady state) by concentration factors, which are the ratios between the concentration in the organism and in the air or water.

But, if all the above mentioned processes have to be taken into consideration for just a few organisms, the model would easily become too complex, contain too many parameters to calibrate and require more detailed knowledge than it is possible to provide. Often, we even do not have all the relations needed for a very detailed model, as toxicology and ecotoxicology are still in their infancy or, said differently, there are too many processes and effects of a big number of different chemical compounds, and we cannot know them all. Therefore, most models in this class will not consider too many details of the partition of the toxic substances in organisms and their corresponding effects, but rather be limited to the simple accumulation in the organisms and their effects. Usually, accumulation is rather easy to model and the following simple equation is often sufficiently accurate:

$$dC/dt = (ef \times Cf \times F + em \times Cm \times V)/W - Ex \times C = (UPT)/W - Ex \times C \qquad (5.3)$$

where C is the concentration of the toxic substance in the organism; ef and em are the efficiencies for the uptake from the food and medium respectively (water or air); Cf and Cm are the concentrations of the toxic substance in the food and medium respectively; F is the amount of food uptake per day; V is the volume of water or air taken up per day; W is the body weight either as dry or wet matter and Ex is the excretion coefficient (1/day). As can be seen from the equation, UPT covers the total uptake of toxic substance per day by the organism.

The equation has a numerical solution, and the corresponding plot is shown in Fig. 5.4 :

$$C/C_{max} = (UPT \times (1 - \exp(-Ex \times t)))/(W \times Ex) \qquad (5.4)$$

where C_{max} is the steady state value of C:

$$C_{max} = UPT/(W \times Ex) \qquad (5.5)$$

Synergistic and antagonistic effects have not been touched on so far. They are rarely considered in this type of model for the simple reason that we do not have much knowledge about these

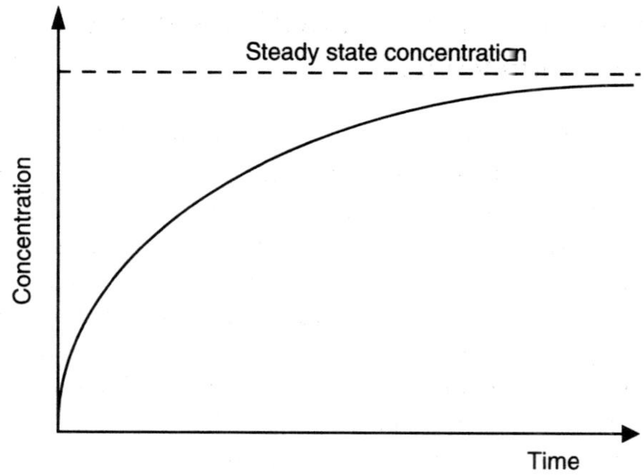

Figure 5.4: Concentration of a toxic substance in an organism versus time (reproduced from [2]).

effects. If we have to model combined effects of two or more toxic substances, we can only assume additive effects, unless we can provide empirical relationships for the combined effect.

Illustration 5.1

An average European is daily inhaling $20\,m^3$ of air with a concentration of $1.5\,\mu g$ cadmium/m^3. The uptake efficiency of metals via the lungs is 50%. The total amount of cadmium in the food for an average European is 300 μg/day. The uptake efficiency for metals via the intestines is 8%. The excretion coefficient for cadmium is 0.001 1/24 h for mammals. What is the steady state concentration and how close to this concentration will a fifty-year-old average European be? The average weight of an average European is 78 kg.

Solution

The daily uptake is:

$$(0.5 \times 1.5 \times 20 + 300 \times 0.08)\,\mu g/24\,h = 39\,\mu g/24\,h$$

The steady state concentration (C_{max}) is $39\,\mu g/(78 \times 0.001) = 500\,\mu g/kg$

The concentration (C) after 50 years of exposure is found by use of eqn (5.4):

$$C/C_{max} = 39(1 - \exp(-0.001 \times 50 \times 365.25)/(78 \times 0.001)/500 = 0.999999998$$

As previously mentioned, estimation methods are widely used in ecotoxicological modelling. The next section will give a brief introduction to the available estimation methods. Ecotoxicological models are used increasingly in ERA and for concrete cases where emission of toxic substances may be a threat for the water-, soil- or air-quality. They have the advantage that they are focusing on the specific problem of a toxic substance, which also defines their applications. If our knowledge about processes and parameters are too limited, the problem is, of course, not solved by another model type. Toxic substances modelling require ecotoxicological models of one class or another, as described in this section.

5.3 The application of parameter estimation methods in ecotoxicological modelling

We are using about 100,000 chemicals in the industrialized countries in such an amount that the chemicals may threaten the environment. We have only sufficient knowledge to make ecotoxicological models with all the needed details for about 500 of these chemicals. Sufficient knowledge means that we know the physical and chemical properties of the chemical (water solubility, octanol-water distribution coefficient, Henry's constant, adsorption constant to soil (100% organic matter is presumed), evaporation rate and rate constant for hydrolysis and photolysis) and the biological properties for a number of organisms (biodegradation rate, biological concentration factor (BCF), biomagnification factor, uptake rate with food and air/water, excretion rate, uptake efficiency, lethal concentration (LC) and lethal dose (LD) values). If we consider, let us say, that about 10 biological properties of a chemical should be known for 1000 different species representing about 10^7 species on earth, to make a proper ERA for the use of the chemical, 100,000 chemicals would require knowledge about $10^5 \times 10 \times 10^3 = 10^9$ parameters. It will be very time consuming to determine so many parameters by *in situ* or laboratory measurements. With the present rate of determining new parameters for the 100,000 chemicals, it will take several hundred years to be able to complete the list of properties. It is therefore obvious that we need estimation methods. Consequently, numerous estimation methods have been developed. The estimation methods have all an uncertainty, which exceeds the uncertainty associated with laboratory determination of the parameters (see the discussion in the book by Jørgensen *et al.* [1]); but it is better to know a parameter with, for instance, an uncertainty of 50% than to know nothing about the parameter. With a knowledge or estimation of the uncertainty it is possible to take the uncertainty into account in the ERA.

Two types of estimation methods are available: specific methods, which are valid for a specific group of chemicals and general methods, which are used for all organic or inorganic compounds. The specific methods have, of course, a more limited application, but have generally less uncertainty. The journal *SAR and QSAR in Environmental Research* contains numerous specific estimation methods. An example to illustrate the specific methods is presented below. The following relationship between 50% growth inhibitory concentration IGC_{50} and the octanol-water distribution coefficient K_{ow} is found valid for phenols in Schultz *et al.* [4]:

$$\text{Log } IGC_{50}^{-1} = 0.627 (\log K_{ow}) - 0.772 \tag{5.6}$$

With $r^2 = 0.783$. The equation is based on 54 phenols and it is assumed that it can be applied on all other phenols. It is just one example out of the several thousands, which can be found in the literature.

A general estimation is possible with the software called estimation of ecotoxicological parameters (EEP) that can be purchased on the home page www.ecologicalmodel.net. EEP applies a network of estimation methods (see Fig. 5.5). It implies, of course, that the uncertainty to estimate, for instance, parameter D through estimations from a known parameter A to an estimated parameter B and to an estimated parameter C, to the estimation of the parameter D is based on the contribution from the uncertainty of all three estimation methods.

If more parameters in the network are known, the uncertainty is of course reduced. A general method is always more uncertain, because the variability of properties among a chemical in general is of course bigger than among chemicals from a specific group of chemicals. The general estimation methods that are applied in a network in EEP should, therefore, preferably be used to give a first more uncertain estimation and in educational context.

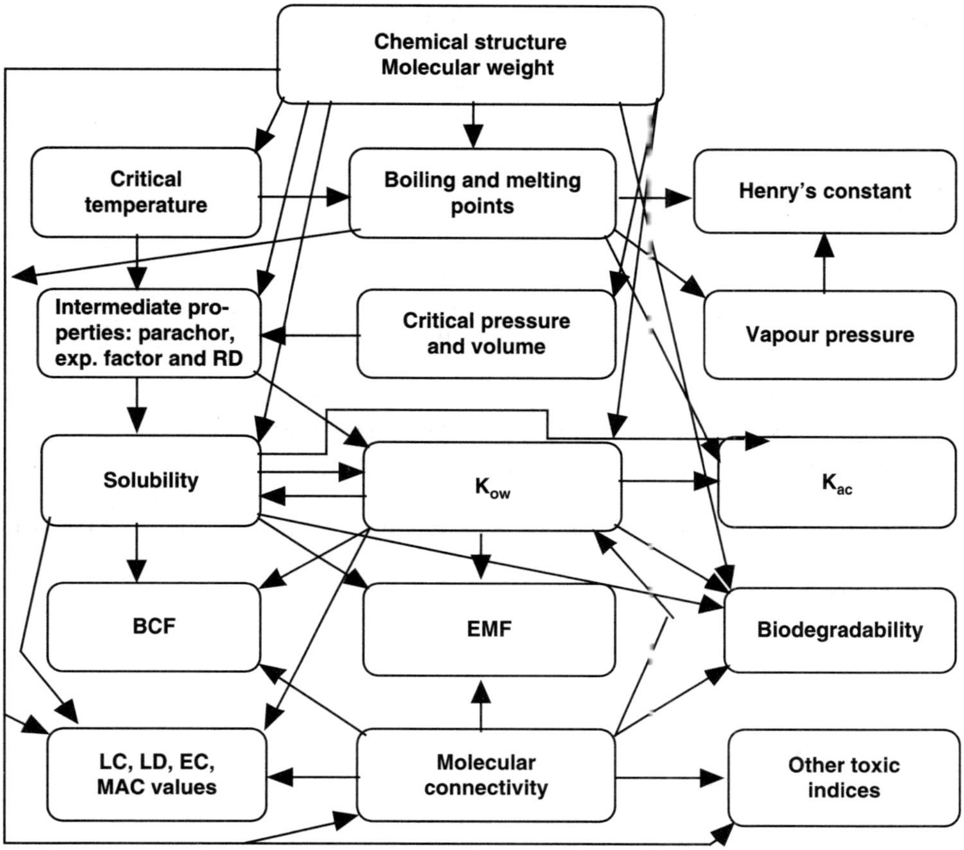

Figure 5.5: A network of estimation methods are shown (see Jørgensen *et al.*, 1998). It illustrates the possibilities of using general estimation methods, but as discussed in the text, the general estimation methods have a higher uncertainty. BCF, EMF the ecological magnification factor and K_{ac} the concentration in soil with 100% organic matter relative to the concentration in water in equilibrium with the soil (reproduced from [2]).

The advantage of EEP is that it is fast to use but the disadvantage is that the uncertainty in some cases is high. If an ERA has to be developed for a chemical, it is necessary to use all accessible estimation methods simultaneously and probably also for a few important parameters to determine them in the laboratory. Consequently, it is recommended to apply EEP to get a first rough estimation of the ecotoxicological parameters or as an educational tool for students to get a fast overview of the wide range of different ecotoxicological parameters. If the uncertainty obtained by EEP is too high, it is recommended to apply more specific estimation methods or to determine the most crucial parameters in the laboratory.

For the development of ecotoxicological models it is important to estimate the uncertainty of the parameters what ever they are based on – parameter estimation methods or measurements to be able to approximate the uncertainty of the model results. In this context, it may be useful to get estimations of all the parameters by EEP and use a comparison of these results with other methods applied for determinations of the parameters to obtain approximate uncertainties.

It is recommendable to get as many different determinations including literature values and values estimated by EEP or other estimation tools for the two most crucial parameters: K_{ow} and the biodegradability. High K_{ow} values tell us that the compounds are soluble in fat tissue, which implies that it will bioaccumulate and biomagnify. If K_{ow} is low, the compounds are readily soluble in water and it is therefore mobile. It means that it will easily be disseminated to adjacent ecosystems. If the compounds are harmful, it will inevitably be difficult to control the effect of a compound with a high water solubility.

A compound with a high biodegradability is, however, decomposed before it may cause ecotoxicological problems by bioaccumulation, biomagnification and high mobility. If the compound on the other hand has a low biodegradability it will stay for long time in the environment (e.g., DDT or polychlorinated biphenyl (PCB)) and if it at the same time has either a high or a low K_{ow}, it will easily cause ecotoxicological problems.

As illustrated by this discussion, with the knowledge about K_{ow} and the biodegradability it is possible to get a first rough estimation of the possible harmful ecotoxicological effects.

Two illustrations of general estimation methods are shown in Figs 5.6 and 5.7; both are reproduced from [2]. Notice (see Fig. 5.7) that mussels and fish have different relationships, which is due to the difference in the relative surface. The surface determines the exposure to the environment in accordance with the allometric principle. The surface is proportional to the weight in the exponent 2/3. The specific exposure becomes therefore proportional to the weight in the exponent – 1/3 or to the length in – 1.

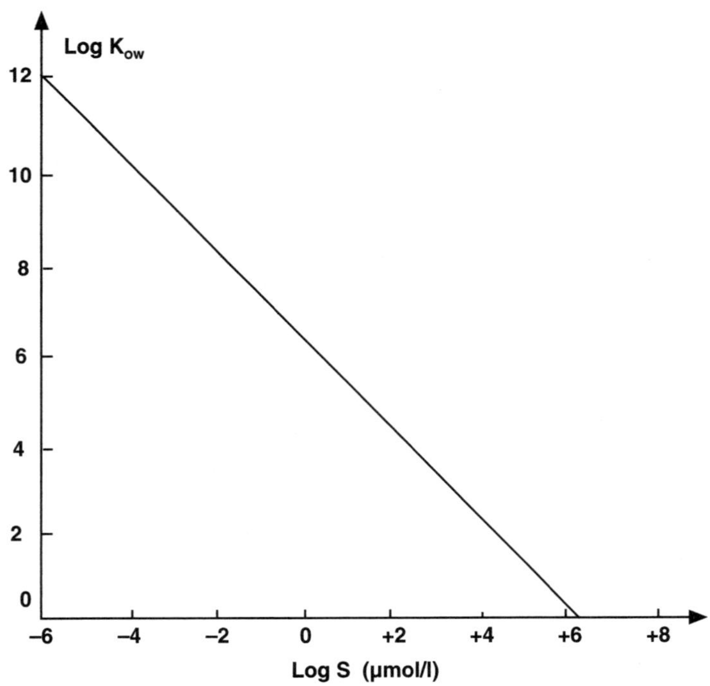

Figure 5.6: A typical general estimation method is illustrated. The octanol-water distribution coefficient (log K_{ow}) is plotted versus the water solubility (log S) in μmol/l. The regression is based on a wide range of organic compounds (reproduced from [2]).

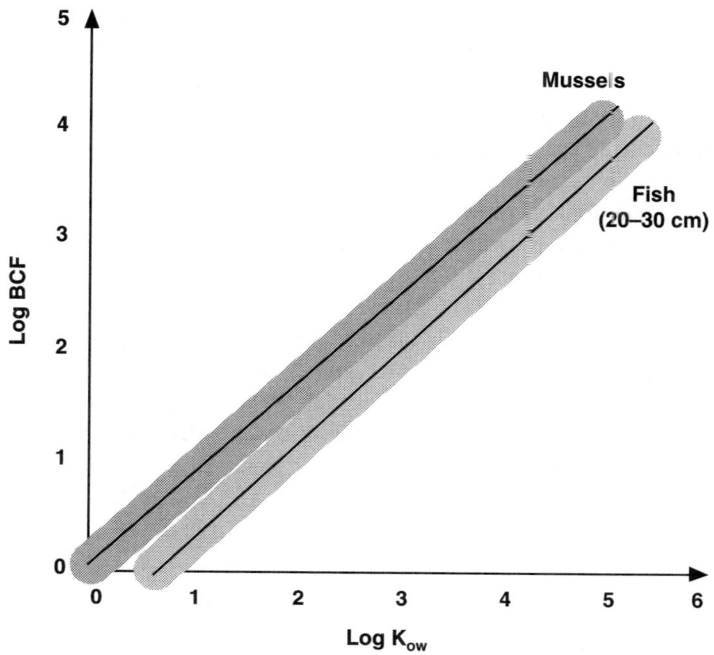

Figure 5.7: The relationship between log BCF for fish 20–30 cm long and mussels (about 4–6 cm) and log K_{ow}. The grey zones around the straight lines indicate the bands corresponding to 95% of the observations of which the correlation is based on. The correlations are based on 142 observations for fish and 82 for mussels. The correlations are based on several sources, of which the most important is Geyer *et al.* [3].

The difference between the two plots in Fig. 5.7 corresponds to a factor 4 (log 4 = 0.6) which is completely in accordance with the allometric principles. Mussels have a length of about 4–6 cm and fish as indicated on the figure about 20–30 cm. Fish with a size of 2–3 m will correspondingly have BCF values – that in the log–log diagram is one unit – lower than the line for 20–30 cm fish, which means that they have BCF values ten times lower.

Illustration 5.2

A chemical compound with a molecular weight of 450 g has a water solubility of 750 µg/l. Estimate the BCF with indication of an approximate uncertainty for a shark with a length of m.

Solution

Figure 5.6 is used to translate the water solubility to K_{ow}. The water solubility in µmol/l is 750/450 = 1.67 µmol/l. Log 1.67 = 0.222, which according to Fig. 5.6 corresponds to log K_{ow} = 6.0.

In Fig. 5.7, it can be found that log BCF for 20–30 cm fish is about 4.0 +/– 0.3 and log K_{ow} = 6.0. It means that BCF is between 5000 and 20,000. The 4 m shark will have a BCF value that is 16 times lower or about 310–1250.

Generally, it is beneficial to apply allometric principles for the estimation of unknown parameters. The surface of an organism is the contact area between the organism and the environment. The rate of exchange between an organism and the environment is therefore proportional to the surface area of the organism or proportional to the volume in the exponent 2/3 or the weight in 2/3 or the length in the exponent 2. Specific rates, it means rate relative to the weight, will therefore be proportional to weight in the exponent – 1/3 or the length in – 1.

Figures 5.8–5.11 give four examples of application of allometric principles. Figure 5.8 a log–log graph giving the specific growth rate as function of the weight. As unicellular organisms, heterotherms and homeotherms have different metabolism, the graph has three lines corresponding to the three different metabolic rates. Figure 5.9 gives the respiration rate as function of the weight. The respiration rate is proportional to the weight in the exponent approximately 2/3 (sometimes the exponent 3/4 may be applied). Figure 5.10 shows the application of the allometric principles for an ecotoxicological parameter. The graph gives the specific uptake rate of cadmium in mg/g 24 h versus the organism's length. Figure 5.11 gives the BCF for cadmium, that is, the ratio of cadmium in water and in the organisms living in the water. BCF has no unit and it is presumed that the concentration in water is mg/liter or kg of water and in the organisms mg/kg wet weight.

5.4 Biogeochemical and ecotoxicological model example: contamination of an agricultural field

The model [5] focuses on contamination of an agricultural field. The model has been calibrated and validated for two antibiotics, but it can in principle be applied for any toxic substances provided, of course, that the processes that are described by the models are valid for the considered

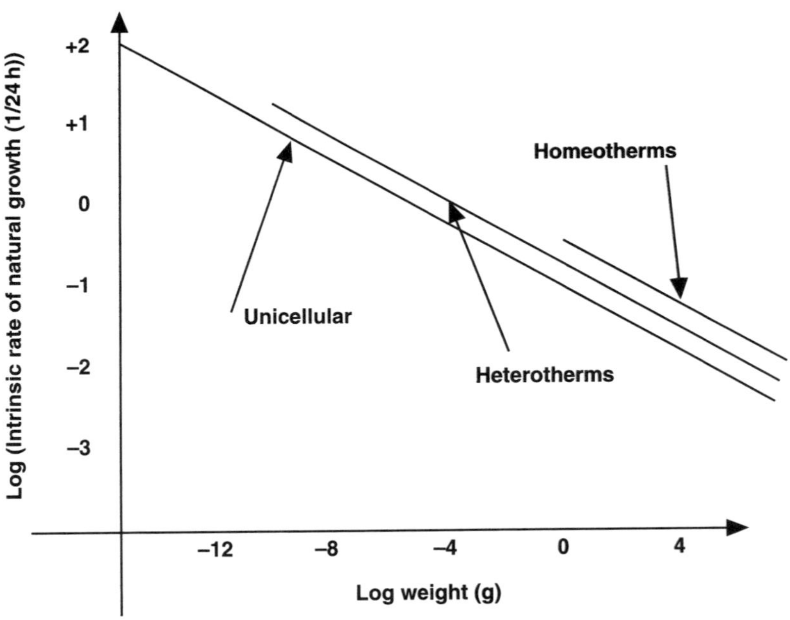

Figure 5.8: The specific intrinsic growth rate versus the weight in a log–log diagram.

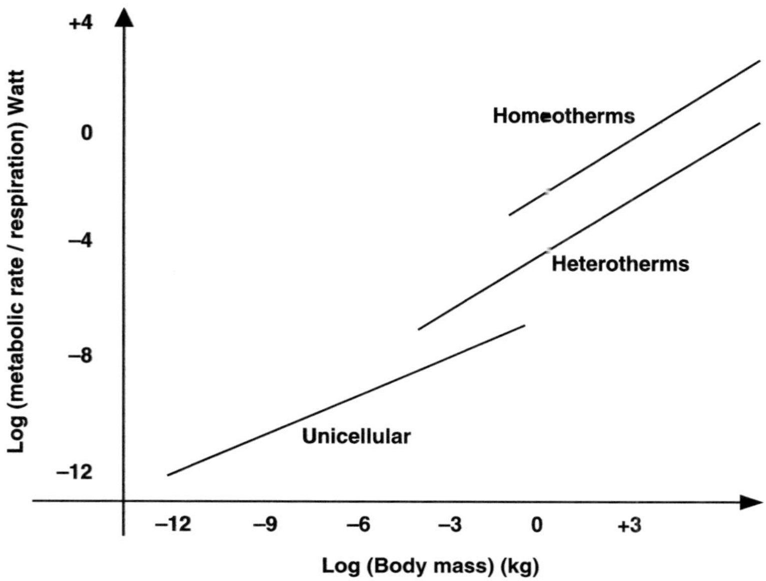

Figure 5.9: The respiration rate in Watt is plotted versus the body mass in a log–log diagram.

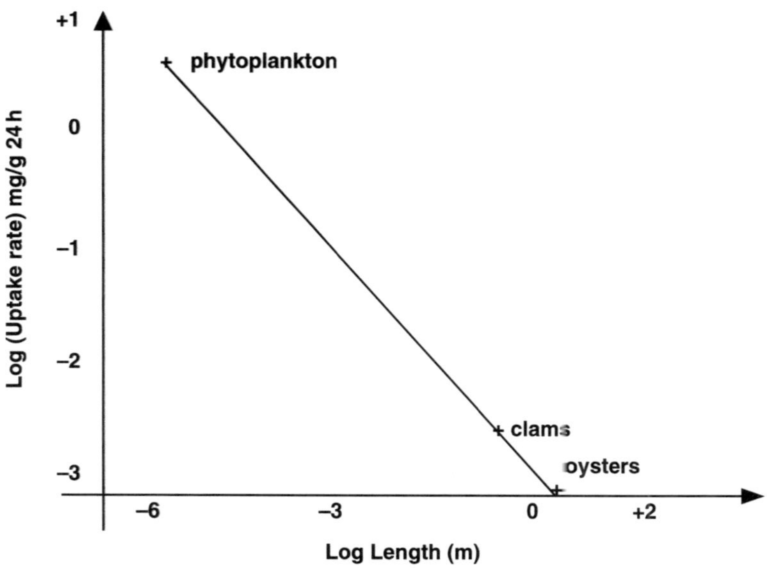

Figure 5.10: The specific uptake rate of cadmium from water as function of the organism's length. The line has the slope – 1, as the allometric theory indicates.

toxic substances. It is often but not always the case. If other processes should be used to describe the dissipation of the toxic substances, it is of course necessary to develop another or slightly revised model. If the processes are valid for another compound, it is still necessary to change the parameters in accordance with the properties of the considered compound, found in the literature,

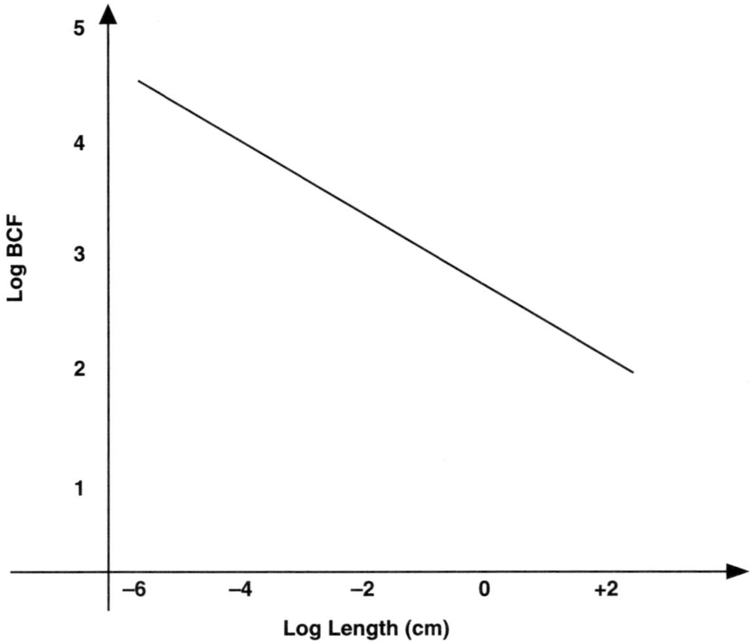

Figure 5.11: The BCF for cadmium as function of length. The plot is linear with the slope – 1. The plot is based on the BCF value for phytoplankotn, zooplankton, shrimps, mussels and goldfish.

by measurements or by application of estimation methods, for instance, EEP. The model described below is for the antibiotic tylosine. The model should, anyhow, always be calibrated and validated for every toxic substance that may be applied on the model.

A software that covers the same processes as for the model presented in Fig. 5.12 is described below in details and uptake by plants is available on the home page www.ecologicalmodel.net. The conceptual diagram for this model with six state variables (toxic substance in plants is added as state variables to the five state variables in Fig. 5.12) is shown in Exercise 5, Chapter 1.

The parameters for both models are, not surprisingly, strongly dependent on the toxic substance considered in the model. Some first estimation of the parameters can be found by use of the software EEP (see Section 5.3). Such fundamental properties as water solubility, K_{ow}, adsorption isotherms and biodegradability are, of course, strongly substance dependent.

The model has five state variables:

- toxic substance on soil water (tsw), μg/l;
- toxic substance in soil (tss), μg/kg;
- metabolites of the toxic substance in soil water (msw), μg/l;
- metabolites of the toxic substance in soil (mss), μg/kg;
- water in the root zone (wrz), l/m^3.

The model has five forcing functions: (notice that capital letters are used for the symbols)

- input of toxic substance to the soil water as function of time (ISW), μg/l 24 h;
- input of the metabolites to the soil water as function of time (IMW), μg/kg 24 h;

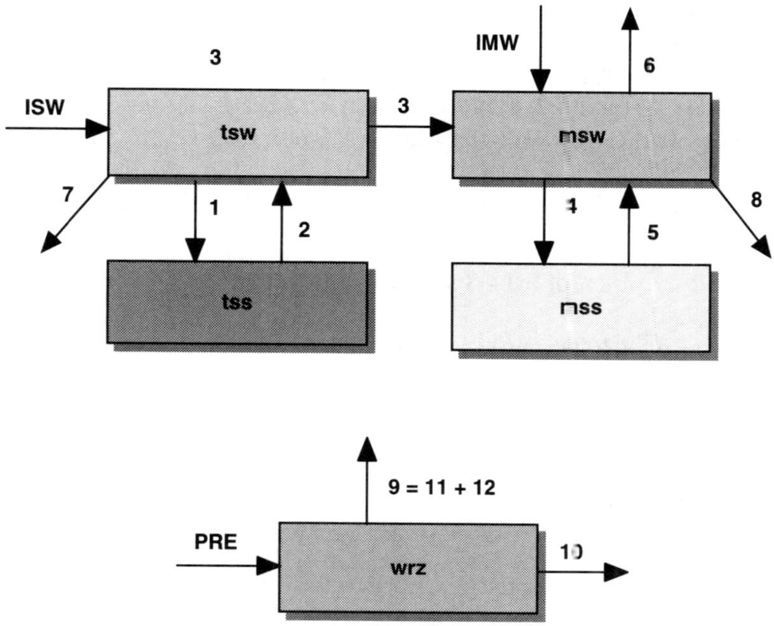

Figure 5.12: Conceptual diagram of the model. The processes are indicated with numbers; see below the equations of the processes: covering adsorption, desorption, decomposition, drainage with the drainage water and evaporation. The forcing functions: precipitation (PRE), input of the toxic substance (ISW) and input of metabolites of the toxic substances (IMW) are directly used in the conceptual diagram.

- precipitation (PRE), mm/24 h;
- radiation (RAD), MJ/24 h;
- temperature (TEM), °C.

Processes

The conceptual diagram is shown in Figure 5.12. The processes 1–12 shown below are described in the model. The numbers are indicated on the diagram, Fig. 5.12. Abbreviation symbols are applied in the equations for the processes (notice that the processes are indicated with three letters and the parameters are indicated with two letters):

1. ads = tsw × ac – tss
2. dsp = tss/ac – tsw
3. dcp = dr × tsw × $1.05^{(TEM-20)}$
4. adm = msw × ad – mss
5. dsm = mss/ad – msw
6. dcm = dm × mss × $1.05^{(TEM-20)}$
7. drt = tsw × dra

8. $drm = msw \times dra$
9. $eva = eps + epp$
10. dra = if wrz >= wf × 400 then (0.1 × (wrz – 400 × wf) else (sl × (400 × wf – wrz)/(ws – wp)
11. $epp = (fa \times la \times 0.931 \times dl \times RAD) /((dl + 66.7) \times 2.47)$
12. $eps = 0.7 \times dl \times RAD \times fa \times \exp(-0.4 \times la) / ((dl + 66.7) \times 2.47)$

The equations have several help-parameters which are calculated from the basic parameters or forcing functions:

$$dl = 3 + 3 \times TEM + 0.17 \times TEM^2$$

fa = if wrz >= wf × 400 then (0.1 × (wrz – 400wf) else ((400 × wf – wrz)/(400wf)

LAI = Lm/(1 + exp(1 – 0.01 × TIME) (TIME is in days, which is the time unit). La is the leaf area index it means the leaf area in m^2 per m^2 area. Lm is LAI just before harvest. Notice it should be possible to select between the application of calculations for LAI by the shown equation or to give information about LAI as function of time. Notice that the time applicable for the shown equation is the growing season and if there are no plants it is a matter of setting Lm = 0.

The water retention capacity of soil is described by three parameters, wf, wp and wxs, describing roughly the minimum water that is retained, the normal water retention and the maximum water retention capapcity. The three parameters can be found from the composition of the soil (cl is clay ratio, si silt ratio and oc organic carbon ratio in the soil):

$$wf = 0.46 \times cl + 0.305 \times si + 0.25 \times oc$$

$$wp = 0.33 \times cl + 0.12 \times si + 1.6 \times oc$$

$$wxs = 0.69 \times cl + 0.55 \times si + 4.28 \times oc$$

Parameters

The basic parameters are shown in Table 5.1 together with the possible ranges and the recommended default values, that are taken from the application of the model on tylosine. These parameter values have been determined by calibration and validation of the model using observed data. The model is included in a CD as a ready to use software coded in C++ in the Handbook of Ecological Modelling and Informatics by Jørgensen, Chon and Recknagel (2009); see also reference [12] Chapter 1 p.25.

The parameters in bold cannot be changed in this software, as they are determined by the characteristic properties of tylosine. They are only valid for the case study "Tylosine". The soil properties and the slope can however be changed in accordance with the focal case.

Differential equations

The differential equations (as coded in C++):

$$dtsw/dt = isw - ads + dsp - dcp - drt$$

$$dtss/dt= iss - dsp + ads$$

$$dmsw/dt = dcp - drm - adm + dsm$$

Table 5.1: The basic parameters are.

The parameter	The symbol used	Unit	Range	Default value
Slope	sl	–	0.00–01	0
Clay	cl	–	0.001–02	0.113
Silt	si	–	0.001–0.2	0.107
Organic C	**oc**	**–**	**0.000–0.2**	**0.016**
Adsorption coefficient	**ac**	**kg/l**	**0–10,000**	**128**
Adsorption coefficient for metabolites	**ad**	**kg/l**	**0–10,000**	**128**
Decomposition rate	**dr**	**1/24 h**	**0–100**	**0.35**
Decomposition rate For metabolites	**dm**	**1/24 h**	**0–100**	**0.20**

The parameters in bold have been found valid for the case study tylosine. The soil properties and the slope can however be changed in accordance with a specific case study.

$$\text{dmss}/\text{dt} = \text{adm} - \text{dsm}$$

$$\text{dwrz}/\text{dt} = \text{PRE} - \text{eva} - \text{dra}$$

Usually it is not possible to observed tsw and tss separately and msw and mss separately. Therefore two additional state variables are introduced:

$$\text{tts} = \text{tsw} + \text{tss and}$$

$$\text{tms} = \text{msw} + \text{mss}$$

5.5 Uptake of toxic substances by plants

Plants are able to take up toxic substance dissolved in the soil water. It makes it possible to remove toxic substances by plants from contaminated soil – an ecological engineering method, which belongs to the group of methods called phytoremediation. It is on the other hand a problem that edible crops would be contaminated by toxic substances, taken up from contaminated soil.

Sludge from waste water treatment contains small concentrations of various toxic organic compounds and heavy metals. There are standards for the concentrations of a few toxic organic compounds and for heavy metal concentrations that can be accepted for sludge applied as soil conditioner in agriculture. The standards reflect the uptake of toxic compounds by plants.

For toxic organic compounds, the following equation for the ratio concentration in plants divided by the concentration in soil, which can be considered as a BCF value for the transfer from soil to plants:

$$\text{BCF} = (1.01\times f_{lip}/f)\ K_{ow}^{0.09} \tag{5.7}$$

where f_{lip} is lipid fraction in the plants and f is the fraction of organic carbon in soil.

For heavy metals, the distribution between soil and soil water determines how much the plants can take up of the total amount of heavy metals. The distribution is determined by the composition of soil: how much clay and humus does the soil contain? and what is the pH and the ion exchange capacity of the soil, CEC? The following equation is valid for cadmium (see [6, 7], where the equations of the distribution for other heavy metals also can be found):

$$\text{distribution for Cd} = 0.00001 \times (800 - 6.14 \times \text{pH}-0.26 \times \text{clay}-0.52 \times \text{humus}-0.93 \times \text{CEC}) \quad (5.8)$$

The uptake of cadmium by plants is dependent on the protein content of the plants (in percentage of dry matter). Figure 5.13 shows a conceptual STELLA diagram for a model focusing on the contamination of plants by cadmium. Notice that the cadmium in soil has two state variables: Cd bound and Cd in soil, which means that the model distinguishes between very firmly bound cadmium and less firmly bound cadmium. A model with only one state variable for cadmium in the soil would not be able to give satisfactory results. The sources for the cadmium contamination is the addition of cadmium to soil due to the use of fertilizers and/or sludge as soil conditioner and the atmospheric fall out mainly by dry deposition.

The equations of the model are shown in Table 5.2, and Fig. 5.14 gives the validation results of the model. The parameters and the calibration were based on pot experiments and the validation was based on large scale experiments adding cadmium containing fertilizers and sludge to the soil at time = 0. The cadmium content in the plants was determined at harvest in the first, second and third year. The following concentrations were found (average of 20 randomly collected maize plants):

1. harvest 1.7 μg/kg dry matter,
2. harvest 1.1 μg/kg dry matter,
3. harvest 0.8 μg/kg dry matter.

These values can be compared with prognosis by the model, respectively 1.7 μg/kg dry matter, 1.6 μg/kg dry matter and 1.2 μg/kg dry matter (see Fig. 5.14). If it is considered that the model is relatively simple – only four state variables – the result of the validation seems acceptable, also because it is an ecotoxicological model where a high deviation, in this case, about 50% at the second and third harvest can be accepted. The predicted values are higher than the observed ones, which means that the model is at least not increasing the risk for cadmium contamination by crops harvested on areas that have received sludge as soil conditioner in addition to the cadmium contamination of the usually applied fertilizers and the general background air pollution on the country side. The concentration of cadmium is, of course, decreasing by the subsequent harvests in the second and third year, because no additional sludge is added to the soil after time 0, but a part of the cadmium is removed by harvest each year and a minor part is also lost to the ground water (indicated in the conceptual diagram as Cd loss). Moreover, the amount of cadmium in detritus will slow down the cycling of cadmium and thereby reduce the amount of easily available cadmium. The cadmium in soil is reduced over the simulation period from about 80 μg/kg dry matter in soil to about 45 μg/kg dry matter in soil (see Fig. 5.14).

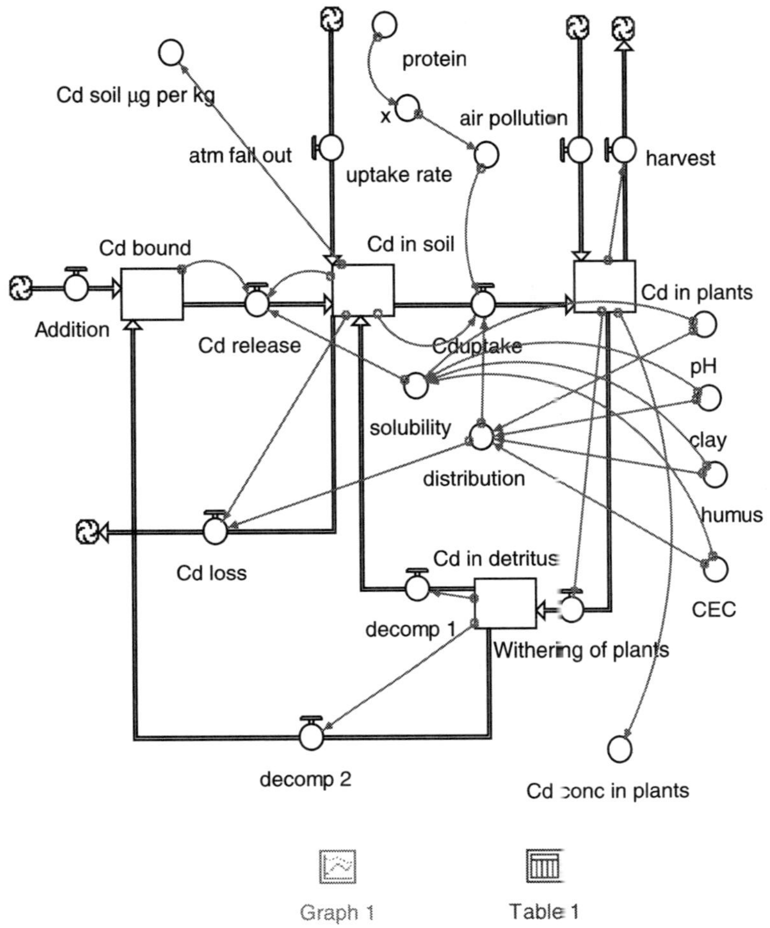

Figure 5.13: STELLA diagram of a cadmium model. The model relates the cadmium in fertilizers and soil conditioners added to the soil and the dry deposition of cadmium on plants and soil with the cadmium concentration in the plants as function of time.

Table 5.2: Model equations (STELLA format).

Cd bound (t) = Cd bound (t – dt) + (decomp2 + Addition – Cd release) × dt
INIT Cd bound = 0.19

INFLOWS:
decomp2 = Cd in detritus × 0.012
Addition = PULSE (0.14, 1, 960)

OUTFLOWS:
Cd release =
IF (Cd in soil < solubility) THEN (0.000012 × Cd bound) ELSE (Cd bound × 0.0000012)
Cd in detritus (t) = Cd in detritus (t – dt) + (Withering of plants – decomp1 – decomp2) × dt
INIT Cd in detritus = 0.27

Table 5.2: Model equations (STELLA format). *(Continued)*

INFLOWS:
Withering of plants = PULSE (0.6 × Cd in plants,180,360) + PULSE (0.6 × Cd in plants,181,360)

OUTFLOWS:
decomp1 = IF (TIME > 180) THEN (0.005 × Cd in detritus) ELSE (0.0001 × Cd in detritus)
decomp2 = Cd in detritus × 0.012
Cd in plants (t) = Cd in plants (t – dt) + (Cd uptake + air pollution – Withering_of_plants – harvest) × dt
INIT Cd in plants = 0.00002

INFLOWS:
Cd uptake = Cd in soil × distribution × uptake rate
air pollution = 0.0000028 + STEP (– 0.0000028,180) + STEP (0.0000028,360) + STEP (– 0.0000028,540) + STEP (0.0000028,720) + STEP (– 0.0000028,900)

OUTFLOWS:
Withering of plants = PULSE (0.6 × Cd in plants, 180,360) + PULSE (0.6 × Cd in plants, 181,360)
harvest = PULSE (0.4 × Cd in plants, 180,360) + PULSE (0.4 × Cd in plants, 181,360)
Cd in soil (t) = Cd in soil (t – dt) + (Cd release + decomp1 + atm fall out – Cd uptake – Cd loss) × dt
INIT Cd in soil = 0.08

INFLOWS:
Cd release = IF (Cd in soil < solubility) THEN (0.000012 × Cd bound) ELSE (Cd bound × 0.0000012)
decomp1 = IF (TIME > 180) THEN (0.005 × Cd in detritus) ELSE (0.0001 × Cd in detritus)
atm fall out = 0.0000014

OUTFLOWS:
Cd uptake = Cd in soil × distribution × uptake rate
Cd loss = 0.1 × Cd in soil × distribution
Cd conc in plants = 5000 × Cd in plants/14
Cd soil microg per kg = Cd in soil × 1000

CEC = 8
clay = 4
distribution = 0.00001 × (800 – 6.14 × pH – 0.26 × clay – 0.52 × humus – 0.93 × CEC)
humus = 2
pH = 7
protein = 47
solubility = 10^(6.273 – 1.505 × pH + 0.00212 × humus + 0.002414 × CEC – 0.00122 × clay) × 112.4 × 350
uptake rate = x + STEP (– x,180) + STEP (x, 360) + STEP (– x, 540) + STEP (x, 720) + STEP (– x, 900)
x = 0.00216 × (– 0.377 + 0.4544 × protein)

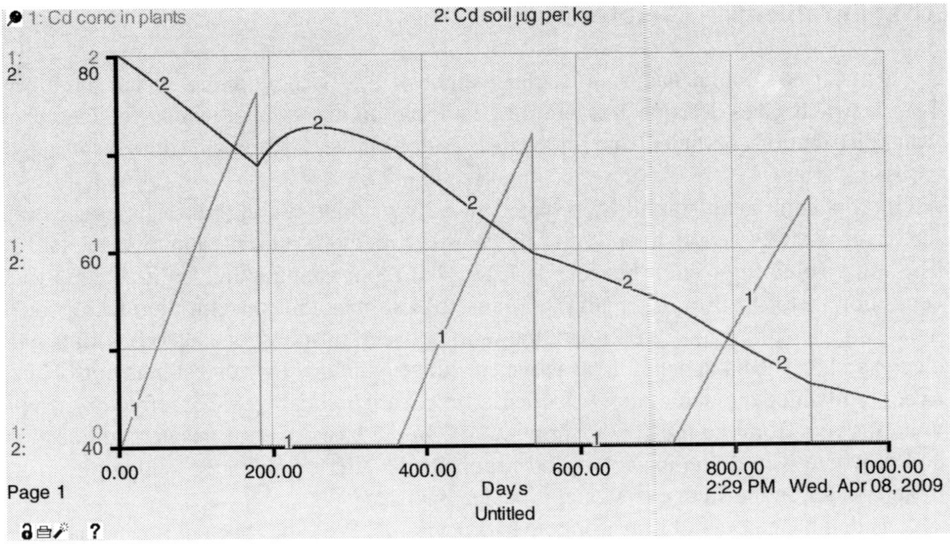

Figure 5.14: The graph shows the cadmium concentration in plants and soil in µg/kg dry matter. The harvest takes place at day 180, 540 and 900. The cadmium concentration according to observations is found at the three harvests to be respectively 1.7, 1.1 and 0.8 µg/kg dry matter. The cadmium in soil is reduced over the simulation period from about 80 µg/kg dry matter in soil to about 45 µg/kg dry matter in soil.

References

[1] Jørgensen, S.E., Halling-Sørensen, B. & Mahler, H., *Handbook of Estimation Methods in Ecotoxicology and Environmental Chemistry*, CRC Press: Boca Raton, FL, 1998.

[2] Jørgensen, S.E. & Bendoricchio, G., *Fundamentals of Ecological Modelling*, 3rd edn, Elsevier: Amsterdam, 2001.

[3] Geyer, H., Sheehan, P., Kotzias, D., Freitag, D. & Korte, F., Estimation of biological concentration factors. *Chemosphere*, **10**, pp. 1307–1320, 1981.

[4] Schultz, T.W., Bearden A.P. & Jaworska, J.S., A novel QSAR approach for estimating toxicity of phenols. *SAR and QSAR in Environmental Research*, **5(2)**, pp. 99–113, 1996.

[5] Halling-Sørensen, B. & Jørgensen, S.E., Modelling the dissipation of two antibiotics tetracyclines and tylosine and their degradation products in soils – a Danish field-case study. *Ecological Modelling*, submitted. The details of the model can also be found in Jørgensen, S.E., Chon, T. & Recknagel, F., *Handbook of Ecological Modelling and Informatics*, WIT Press: Southampton, 2009.

[6] Jørgesen, S.E., Model of heavy metal contamination of crops and ground water. *Ecological Modelling*, **2**, pp. 59–68, 1976.

[7] Jørgensen, S.E., Removal of heavy metals from compost and soil by ecotechnological methods. *Ecological Engineering*, **2**, pp. 89–100, 1993.

Exercises/problems – Chapter 5

1. A chemical compound has a molecular weight of 320 g/mole and a water solubility of 1 mg/l, which gives a vapor pressure of 2 Pa. Find an approximate value for the octanol-water distribution coefficient and for the distribution between water and soil with 4% organic carbon.
2. A chlorophenol has a solubility in water of 1.2 mg/l. Find the approximate concentration factor for mussels (length about 5 cm) by using the methods presented in this chapter.
3. The water solubility of nitrobenzene is 1.27 g/l at room temperature. (a) Estimate the concentration factor for fish of length 20–30 cm. (b) Estimate the concentration factor for blue mussels of length 5 cm. (c) Estimate the ratio between nitrobenzene adsorbed to activated sludge and dissolved in water in an activated sludge plant. (d) Do you estimate nitrobenzene to be readily biodegradable, slowly biodegradable, very biodegradable or refractory? Explain your answers from the molecular formula. (e) On the basis of your answers to questions c and d, where would you expect to find nitrobenzene after an activated sludge plant, in the treated water, in the sludge or would it be decomposed?
4. A considerable amount of polyaromatic hydrocarbons (PAH) has been dumped on a rubbish dump. The environmental department of the city fears that the ground water will be contaminated. (a) Calculate the percentage of PAH adsorbed to the soil and in equilibrium with PAH in soil water. The soil is known to contain 10% carbon. Assume that log $K_{ow} = 5.5$. (b) Henry's constant for the mixture of PAH is 0.002 atm. What is the concentration in air in equilibrium with soil water containing 80 mg/l? Is it expected that PAH will be relatively rapidly removed by evaporation?
5. Explain why the concentration of most micro pollutants in an organism increases with time (and weight of the organism)?
6. DDT has been banned and in many countries at least partly replaced by chlordane. Estimate the difference in physical–chemical properties between the two pesticides based on their chemical structure. What difference is expected in the bioaccumulation and persistence of these two substances? Find the chemical formula on the internet for the two pesticides.
7. What is the expected difference in environmental behavior between phenoxyacetic acid containing the COOH group and the corresponding sodium salt containing COONa? What would be the difference between the two forms in relation to accumulation in soil and evaporation? Use the results to indicate how pH will influence these properties for phenoxyacetic acid?
8. What would be the difference in biodegradability of a branched alkylsulphonate (carbon chain with 12 C-atoms, 7 branches) and a completely linear alkylsulphonate with the same number of carbon atoms? Use the rules presented in the text to indicate the difference semi quantitatively.
9. Log Kow for atrazine is 2.75. Estimate the distribution between soil with 1.8% organic carbon and water for atrazine. What would be the estimated ratio of the concentration in carrots with 0.6% lipid to the concentration in carrots grown in atrazine-contaminated soil?
10. Indicate by an "x" in table shown below in which classes do the eight compounds in the table belong. Class 1 covers the compounds that are decomposed at least 10% in a biological treatment plant and eventually after adaptation significantly more than 10%. Class 2 comprises compounds which are 1–10% biodegraded in a biological treatment plant while class 3 means that the compounds are not biodegraded (< 1%) in a biological treatment plant.

Compound	Class 1	Class 2	Class 3
Ethylenglycol			
1,3 dichloroanthracene			
2,3,4-trinitrophenol			
DDT			
PCB			
Glycerine			
Dioxin			
Pentachlorphenol			

Are the compounds that are slowly biodegraded (class 2 and 3) accumulated in the water phase or in the sludge phase?

11. Hexachlorobenzene has an octanol-water partition coefficient of $10^{6.18}$. Find the approximate concentration factor for 25–30 cm length fish. Find also BCF for shrimps (length approximately 4 cm) presuming the same percentage fat tissue. Find finally the concentration of hexachlorobenzene in soya beans with a lipid content of 8.5% cultivated in soil with 2% organic carbon and with a concentration of hexachlorobenzene of 12 mg/kg dry matter.

CHAPTER 6

Spatial models

6.1 Introduction

Forcing functions and parameters representing the properties of the organisms in ecosystems may often be spatially distributed. It implies that the state variables, which are considered as the result of ecological models also may be spatially distributed. Spatial distribution is characteristic for most ecosystems and it would therefore be important to be able to develop ecological models with spatial distribution. The question that will be discussed in this chapter is: how is it possible to develop spatial models?

Four methods to deal with the spatial distribution will be presented in this chapter:

1. The model of a constructed subsurface wetland presented in Figs 3.12–3.14 is based on a model description based on five boxes. The forcing functions corresponding to the water quality of the inflow water, the temperature of the water and the parameters, are not spatially distributed, but the resulting water quality, that is, the BOD_5, the ammonium-N, the nitrate N, the organic-N and the phosphorus concentrations (all state variables in mg/l) are all different in the five boxes, because what is the resulting water quality in box 1 is the inflowing water quality for box 2 and so on until the resulting treated water flowing out of box 5 gives the final result of the treatment by the subsurface wetland. The model is a one-dimensional (1D) model because it is just five sequential boxes. A 1D river model could be developed in a similar way. It is of course possible to divide an area or a volume into boxes and thereby build two-dimensional (2D) and three-dimensional (3D) models. The next section will present this approach.
2. It is also possible as it will be illustrated in the Section 6.3 to use distance as the independent variable instead of time.
3. Section 6.4 will show how it is possible to present spatial results by use of geographical information system (GIS). Most available GIS software makes it even possible to link GIS with model calculations. It is named surface modelling and it will shortly be mentioned in Section 6.6.
4. Section 6.5 will demonstrate how the modelling tool cellular automata or lattice can be used to build spatial models. The lattice approach can be used to represent the spatial distribution, but it can also be used to represent individuals of a population. In the first case, we want to emphasize the spatial distribution and therefore call the resulting model a spatial model, while we will call the resulting model an individual based model when the differences among the individuals of a population are emphasized. In the last section (Section 6.7) the lattice approach is used to illustrate an individual based model (IBM), although the same model could in principle also be considered as a spatial model, but in this case the individual differences are

considered as the most important feature of the model. The model focuses on the role of conjugation in the evolution.

6.2 2D and 3D models

The classical method to account for the spatial distribution is to divide the ecosystem into boxes. It is, of course, possible to take into consideration that the forcing functions and the parameters (properties) are different in the different boxes.

2D and 3D models can be considered as an approach to solution of partial differential equations, where the boxes discretize the spatial continuum. Figure 6.1 shows a 2D spatial model, where nine boxes are applied to describe the spatial differences. For all the nine boxes, models are developed with forcing functions, parameters and state variables. In addition, the nine models are interconnected as shown in the figure. Transfers of mass, energy and information from one box to all the adjacent boxes may take place. Quantitative descriptions of all these transfer processes are needed for the development of a model for the entire ecosystem, but it implies of course increased complexity of the model. For transfer of mass, it is for aquatic ecosystems a question about the hydrodynamics. As the hydrodynamics is a much faster process than the biological processes in aquatic ecosystems, the hydrodynamic transfer processes are described by much shorter time steps than the biological processes. For terrestrial ecosystems, the transfer processes are usually slower than for aquatic ecosystems but the transfer processes are complicated by a high heterogeneity.

2D and 3D models, because of their higher complexity are more time consuming to develop particularly for the calibration and validation. Moreover, they will require more computer power. It is therefore not surprising that most 3D models have been developed during the last decade (see, for instance, the 3D lake model by Hu *et al.* [1]).

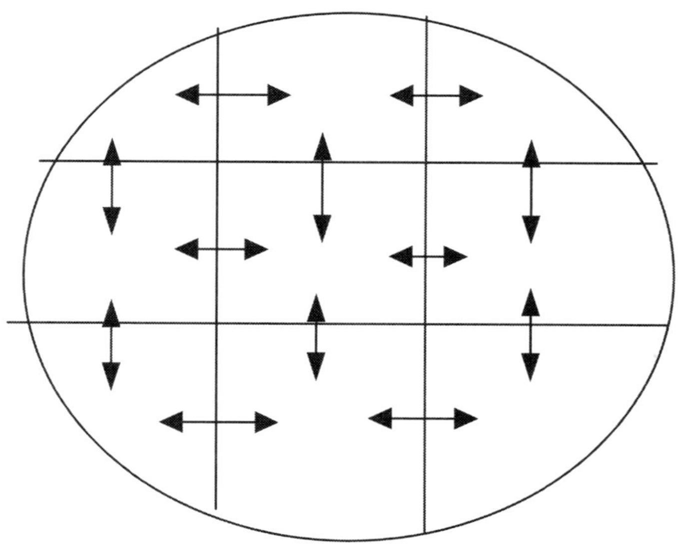

Figure 6.1: A 2D model of an ecosystem is shown. The ecosystem is divided into nine boxes and the transfer to adjacent boxes have, of course, to be considered in the model description.

6.3 An example of distance as independent variable: chromium pollution in a Danish Fjord

This case study combines an ecotoxicological fate model type (a specific ecosystem is considered) with an effect model. The structure of the model is simplified by focusing on a steady state situation, although the spatial distribution is considered. Only one trophic level is considered. It is, however, an illustrative case study, because:

1. The objectives of the model are achieved by a simple model.
2. It has been possible to validate the prognosis 8 years after it was published. Validations of models are not only important but also absolutely necessary for the development of reliable models. Here it has been even possible to validate the model predictions. Unfortunately, we have only very few cases of prognosis validations. Therefore, it has been considered significant to include this case study.
3. The model development shows clearly, how important it is to know the system and its processes, if the right model with the right simplifications is to be selected.
4. Distance is used as an independent variable instead of time.

A map of the system, Faaborg Fjord, is shown in Fig. 6.2. Faaborg Town (By) is situated on the Danish island Fyn. The numbers show sampling stations. Station 1 is of particular importance, as it is close to the discharge point.

A tanning plant has discharged wastewater for decades with a high concentration of chromium (lll) into the fjord. In 1958, production was expanded significantly and gave a pronounced increase in the chromium concentration of the sediment [2].

It was the scope of this investigation to set up a model for the distribution of chromium in the fjord based on the analysis of chromium in phytoplankton, zooplankton, fish, benthic fauna, water

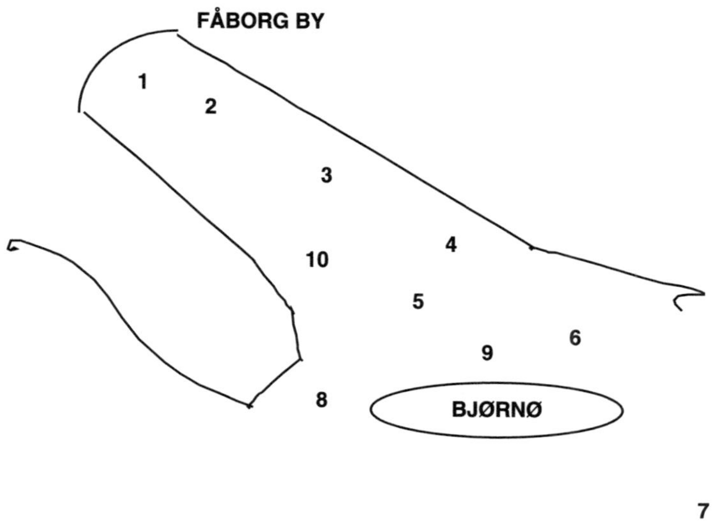

Figure 6.2: Faaborg Fjord. The sampling stations 1–10 are shown. The point close to sampling station 1 indicates the discharge point.

(dissolved as well as suspended) and sediment. Already during the first phase of the investigation, it was clear that the phytoplankton, zooplankton and fish were almost not contaminated by chromium, while the sediment and the benthic fauna clearly showed raised concentrations of chromium. This was easy to explain: the chromium (III) precipitates as hydroxide by contact with the seawater that has a pH of 8.1 compared with pH of 6.5–7.0 for the wastewater. The benthic fauna becomes thereby an indicator for the chromium contamination, which, however, is decreasing as the distance from the discharge point is increasing.

6.3.1 Model description

The overall analysis showed that the important processes are:

1. Settling of the precipitated chromium (III) hydroxide and other insoluble chromium compounds.
2. Diffusion of the chromium, mainly as suspended matter, throughout the fjord is caused mainly by tides. It implies that an eddy diffusion coefficient has to be found. Advection may also play a role, but due to the relatively small amount of tanning wastewater, the importance of advection is probably less than diffusion caused by the tide.
3. Bioaccumulation from sediment to benthic fauna.

Processes 1 and 2 can be combined in one submodel, while process 3 requires a separate submodel.

A mathematical description of advection and diffusion processes is combined with a description of settling in accordance with the important processes 1 and 2 mentioned above: the distribution model is based on the following simple chromium (III) transport equation (see, for instance, other similar cases in Weber [3]).

$$\delta C/\delta t = D \times \delta^2 C/\delta X^2 - Q \times \delta C/\delta X - K \times (C - Co)/h \tag{6.1}$$

where C is the concentration of total chromium in water in mg/l, C_o is the solubility of chromium (III) in seawater at pH = 8.1 in mg/l, Q is the inflow to the fjord = outflow by advection (m^3/24 h), D is the eddy diffusion coefficient considering the tide (m^2/24 h), X is the distance from the discharge point in m, K is the settling rate in m/24 h and h is the mean depth in m.

For a tidal fjord such as Faaborg Fjord, the advection Q is as already indicated insignificant compared with the tide and it is possible with good approximations to set Q to 0. The tanning plant has discharged an almost constant amount of chromium (III) during the last two decades. We can therefore consider the stationary situation:

$$\delta C/\delta t = 0 \tag{6.2}$$

eqn (6.1) takes accordingly the form:

$$D \times \delta^2 C/\delta X^2 = K \times (C - Co)/h \tag{6.3}$$

This differential equation of second order has an analytical solution. The total discharge of chromium in g per 24 h, denoted as Cu, is known. This information is used together with F = cross sectional area (m^2) to state the boundary conditions. The following expression is obtained as an analytical solution: (see any mathematical textbook or handbook).

$$C - Co = (Cu/F) \times \sqrt{(h/D \times K)} \times \exp[-\sqrt{(K/h \times D)} \times X] + IK \tag{6.4}$$

F is known only approximately in this equation due to the non-uniform geometry of the fjord. The total annual discharge of chromium is 22,400 kg. Both the consumption of chromium by the tanning factory and the analytical determinations of the wastewater discharged by the factory confirm this number. h, the mean depth, is about 8 m on average. IK is an integration constant.

Equation (6.4) may be transformed to:

$$Y = K \times (C - Co) = (Cu/F) \times \sqrt{(h \times K/D)} \times \exp[-\sqrt{K/h \times D}) \times X + K \times IK \qquad (6.5)$$

Y is, as seen, = K × (C – Co), that is, the amount of chromium (g) settled per 24 h and per m^2. The equation gives, as seen, Y as a function of X. Y is, however, known from the sediment analysis at the 10 stations. A typical chromium profile for a sediment core is shown in Fig. 6.3. As we know that the increase in the chromium concentration took place about 25 years before the model was built, it is possible to find the sediment rate in mm or cm per year: 75 mm/25 y = 3 mm/y. As we also know the concentration of chromium in the sediment, we can calculate the amount of chromium settled per year, or per 24 h and per m^2, and this is Y. The Y values found for the 10 stations by this method are plotted versus X in Fig. 6.4. It is easily seen that the

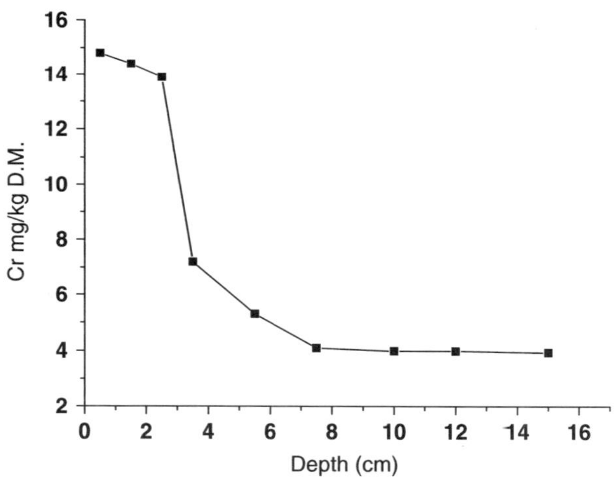

Figure 6.3: Typical chromium profile for a sediment core.

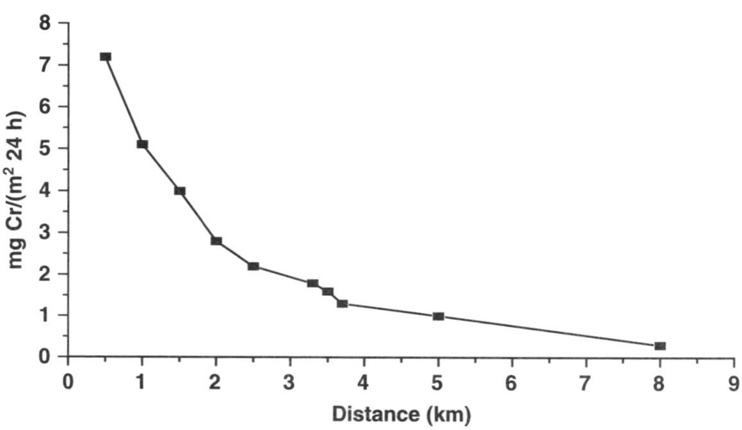

Figure 6.4: Settled amount of chromium as function of the distance from the discharge point.

decrease of settled chromium as function of the distance from the discharge point is close to an exponential decay.

A non-linear regression analysis can be used to fit the data to an equation of the following form:

$$Y = a \times \exp(-bX + c) \tag{6.6}$$

a, b and c are constants, which are found by the regression analysis.

The regression analysis gave the results shown in Table 6.1.

The results of the statistical analysis; see Table 6.2, were convincing as the F-value was found to be 114.5, which gives a probability of more than 99.99% for the exponential equation based on a, b and c.

The settling rates can be found by use of the following equation:

- Y mg Cr/m^2 day = K (C – Co),
- **Y and C – Co are known for all the stations. Co is 0.2 mg/l.**

K is found for the 10 stations. The results are shown for six of the stations in Table 6.3.

Table 6.1 Estimations of a, b and c.

Constants	Estimate	Asymptotic st. error
a	0.009909	0.00084
b	0.000723	0.00015
c	–0.000081	0.00045

Table 6.2 Statistical analysis.

	Degree of freedom	Sum of squares	Mean square
Model	3	0.00011337	0.00003779
Residual	6	0.00000233	0.00000033
Total	9		
	F = 114.5		

Table 6.3 Settling rates.

Station mg	Cr/m^{2}day	$C - C_0$ (mg m^{-3})	K (m day^{-1})
2	6.5	2.5	2.6
4	2.8	1.75	1.6
5	2.1	0.9	2.3
6	0.4	0.6	0.7
7	0.1	0.2	0.5
8	0.6	0.3	2.0

The question is of course if the found K value is reasonable. The average is 1.6 m/24 h, which is reasonable, because of the settling of detritus and phytoplankton – it means organic matter with a specific gravity close to 1.0 g/ml is in the order of 0.05 m–0.5 m /24 h. Chromium hydroxide with a significant higher specific gravity is expected to settle faster.

Integration from 0 to infinite of eqns (6.5) and (6.6) over a half-circle area from 0 to infinite gives a result 22 t of chromium (III), that is, almost all the chromium discharged may be explained by the model, assuming that the distribution takes place over a half-circle area.

Overall, it may be concluded that the distribution model gives acceptable results. The high concentrations of chromium in the sediment give reliable determinations, which again is the basis for the distribution model. The use of sediment analysis, as demonstrated, is recommended for the development of a distribution model for a component as chromium (III) hydroxide that settles readily.

The second submodel focuses on the chromium contamination of the benthic fauna. It may be shown [4] that under steady state conditions the relation between the concentration of a contaminant in the n'th level of the food chain and the corresponding concentration in the (n–1)'th link, which is the food for the n'th level can be expressed by use of the following equation:

$$C_n = (\mu(n) \times C_{n-1} \times Y(n))/(MY(n) \times F(n) - RESP(n) + EXC(n)) = K_c \times C_{n-1} \qquad (6.7)$$

where the symbols have the following definitions and units:

$\mu(n)$ = the maximum growth rate for n'th level of the food chain (1/day).
C_n = the chromium concentration in the n'th level of the food chain (mg/kg).
C_{n-1} = the chromium concentration in the (n–1)'th level of the food chain (mg/kg).
Y(n) = the utility factor of chromium in the food for the n'th level of the food (-).
F(n) = the utility factor of the food in the n'th level of the food chain (-).
RESP(n) = the respiration rate of the n'th level of the food chain (1/day).
EXC(n) = the excretion rate of chromium for the n'th link of the food chain (1/day).
K_c = the accumulation coefficient for the transfer of chromium from the (n–1)'th level to the n'th level of the food chain.

For some species present in Faaborg Fjord, these parameter values can be found in the literature (see, for instance, Jørgensen *et al.* [5]). The mussels *Mytilus edulis* (blue mussels) that were found on almost all the stations were used to find the accumulation from the sediment to benthic animals. The following parameters are valid for *Mytilus edulis* in the literature:

- $\mu(n) = 0.03$ 1/day,
- $Y(n) = 0.07$,
- $F(n) = 0.67$,
- $RESP(n) = 0.001$ 1/day,
- $EXC(n) = 0.04$ 1/day.

The use of these values implies that $K_c = 0.036$ for *Mytilus edulis*. In other words, the concentration of chromium in *Mytilus edulis* should be expected to be 0.036 times the concentration in the sediment. As the concentration of organic matter that is the food for *Mytilus edulis* is in the order of 0.4 ppt, which means that the concentration of chromium in *Mytilus edulis* is about 90 times higher than in the organic matter in the sediment. The major part of the sediment is sand.

Table 6.4 Validation of the prognosis.

Item	mg per kg dry matter		
	Observed value	Range	The prognosis
Cr in sediment	65	57–81	70
Cr in mussels	2.2	1.4–4.5	2.5
mg Cr/m^2 day	0.59	0.44–0.83	0.67

Twenty-one mussels from Faaborg Fjord were analyzed and by statistical analysis, it was found that the relation between the concentration in the sediment and in the mussels is linear:

$$C_n = C_{n-1} \times K_c' \tag{6.8}$$

where K_c' was found to be 0.015 ± 0.002. The discrepancy from the theoretically found value – 0.036 – is fully acceptable, when it is considered that the parameters found in the literature are not exactly the same values for all environments and for all possible conditions. The blue mussels found in Faaborg Fjord have, for instance, high variability in the size. In general, biological parameters can only be considered as approximate values. The relatively low standard deviation of the observed K' value confirms, however, that a linear relationship is valid.

It is proposed to use the highest K_c value = 0.036, when the model is used for environmental management, as in that way the uncertainty of the K' value should be "used to the benefit of the environment".

The model was used as a management tool and the acceptable level of the chromium concentration in the sediment of the most polluted area was assessed to be 70 mg/kg dry matter. That would correspond to a chromium concentration of 70 × 0.036 = 2.5 mg/kg dry biomass in mussels, or about 2.5 times the concentration found in uncontaminated areas of the open sea. This was considered as the NOEC (non-observed effect concentration) and was accepted by the environmental authorities of the district (council).

The distribution model can now be used to assess the total allowable discharge of chromium (kg/y), if the chromium concentration in the sediment should be reduced to 70 mg/kg dry matter in the most polluted areas (stations 1 and 2 about 0.5 km from the discharge point). It was found that the total discharge of chromium should be reduced to a maximum of 2000 kg or less per year to achieve a reduction of about 92%. Consequently, the environmental authorities required the tanning plant to reduce its chromium discharge to ≤ 2000 kg per year. The tanning plant has complied with the standards, since 1982.

A few samples of sediment (6) and mussels (10) were taken in 1988–1989 to validate the model prognosis. The analytical results of the sediment and the mussels are presented in Table 6.4. Settled chromium in mg/m^2 day was calculated on basis of the previous determined sedimentation rate (see above). The prognosis validation was fully acceptable as the deviation between prognosis and observed average values for chromium in mussels is approximately 12%. Based on the successful validation of the prognosis, it can be concluded that the model although it is simple and only consists of two equations that are solved analytically has been able to capture the determining processes. For other metals, it may be necessary to include more processes, because several metals are able to form organic-metal compounds in contrast to chromium. For instance for mercury, which is able to react with the organic matter of the sediment and form volatile methyl– mercury compounds, it has been found necessary to include the reactions between organic matter and mercury to be able to develop a workable model.

6.4 Application of GIS

GIS is widely used in spatial modelling to present the model results and the environmental data applied as forcing functions by a very illustrative method. Typical GIS applications in spatial modelling are:

- Development of a model to answer the important question: Where to construct wetlands for treatment of the drainage water in the landscape to obtain a certain nutrients removal effect ?
- Development of a model to present the spatial distribution of fish, shell fish, prawns etc.
- Development of a forest model to be applied in forest management.
- Application of models in landscape planning.
- Development of coastal zone models to present the distribution of pollutants.
- Development of a model to present the vegetation pattern.

Several types of different GIS software are available in the market. The selection of the best software for a given task requires a careful comparison between what the software can offer and what is needed to give the best illustration of the model results and the forcing functions of the model.

Figure 6.5 presents the result of a model by GIS. The model calculates the nitrogen loss in g/m^2 y. Nitrogen loss is the state variable of the model and the forcing functions are soil properties: clay content, content of organic matter, sand content, hydraulic conductivity, ion exchange capacity and calcium content.

The forcing functions are presented by the same GIS template. The loss of nitrogen is calculated for each of the $50 \times 50\,m^2$ grids by either a causal equation linking the forcing functions and the state variables or an equation found by correlation analysis of the same variables found for another area. The model is linked directly to the GIS to produce the results of the state variables when the spatial distributions for the same grid are given by means of the GIS template. A model development based on GIS is denoted as surface modelling (see Section 6.6).

6.5 Use of cellular automata or lattice models in spatial modelling

We may distinguish between p-states and i-states in structured populations. The p-states represent a whole population, expressed as general and average properties, such as population size, average weight or length etc. The i-states represent states of individual organisms, within the population, such as size, age, sex, growth rate etc. The spaces of the two states are called p-space and i-space. Metz and Dickmann [6] indicate two types of i-space models: i-space configuration models that follow individuals as discrete entities and an i-space distribution approach using continuous functions for instance the size–frequency distribution, f (S), to inform about the size distribution of the individuals of the population.

Lattice models are applied to develop IBMs where the focus is that individuals have different properties and these differences should be taken into account, when the forcing functions – it means the conditions – are changed. According to the two types of models, mentioned above, it is an i-space configuration model.

Software is available for the development of lattice models and use of cellular automata, but the widely used software Mat. Lab also includes this approach.

A lattice model consists of many cells. Rules are formulated to delete, move or add new elements in the cells. Rules are also formulated for other activities, for instance; growth and uptake of food.

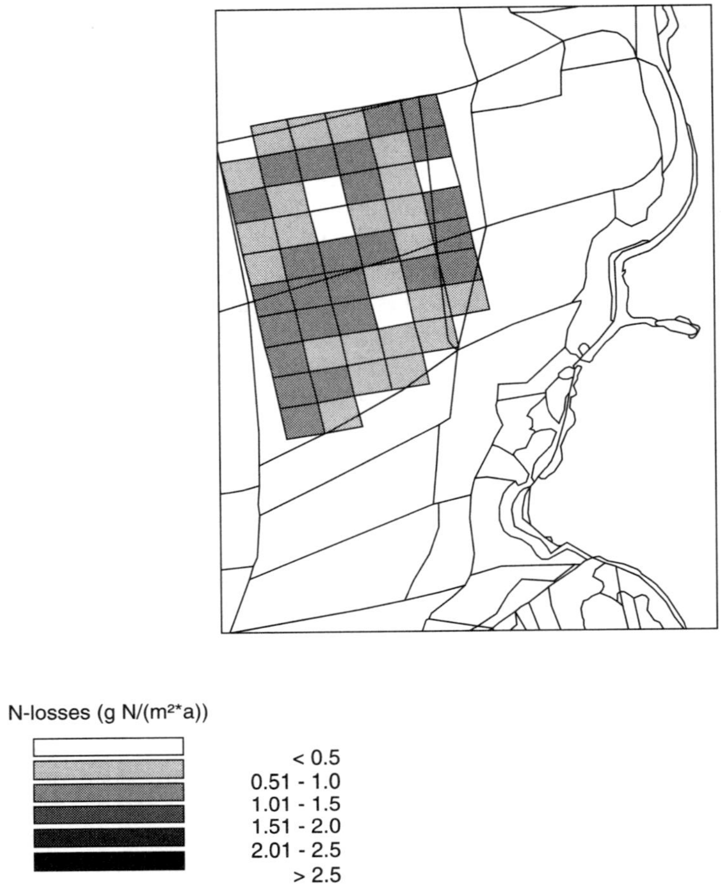

Figure 6.5: The results of a spatial model of nitrogen losses based on the spatial information about the soil properties, which could be presented in a similar way by the use of GIS. The grid scale is 50 m.

The idea to use the lattice model approach to cover spatial distribution or individuality in modelling is illustrated below by a simple example with only few cells (48). The example illustrates the distribution of trees in a forest and how trees are turned over by the wind according to the wind speed, height of the trees and the openings in the forest.

The rules used in this simple illustrative example are:

- THE WIND IN A CELL IS DEPENDENT ON THE NUMBER OF TREES THE WIND WILL PASS AND THE HEIGHT OF THE TREES.
- THE TURN OVER OF A TREE IS DEPENDENT ON THE WIND IN THE INDIVIDUAL CELLS AND THE HEIGHT OF THE TREE.
- NEW TREES BY NATURAL FORMATIONS.
- NEW TREES BY PLANTATIONS.

The result of the model will present the distribution of trees as a film, showing for each cell whether the tree remains, is turned over or a new tree is formed naturally or by plantation. In this

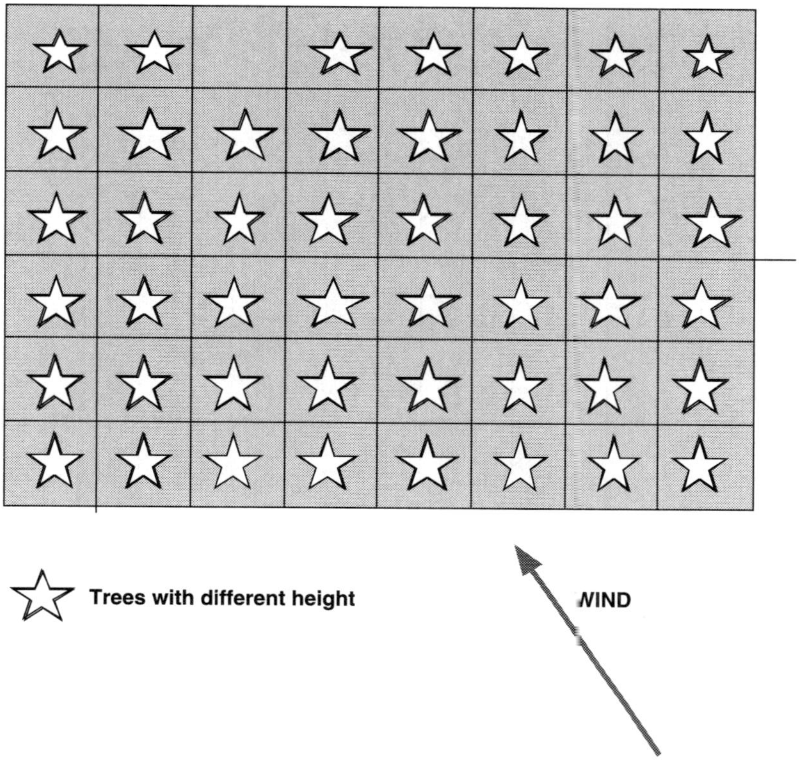

Figure 6.6: Initial stage of a forest model based on the lattice approach.

simple model, the height of the tree increases in accordance with a logistic growth equation. It is possible to introduce factors influencing the growth, for instance, the wind speed, changed fertilizer concentrations etc. Figure 6.6 shows the initial stage of the model and Figs 6.7–6.10 illustrate the development of the model according to the rules and the wind rate as function of time. As seen, the model development is comparable with a film.

6.6 Surface modelling

Surface modelling (SM) results in a numerical representation of the earth's surface (or the surface of another planet) [7] by grids with known coordinates. The model method is a heavy user of GIS.

SM is applied to formulate an ecological object or property in a grid system to represent the spatial distribution. SM was used already in the 1960s, but has greatly progressed since early 1990s due to the rapid development of remote sensing and GIS.

Yue *et al.* [7] present a typical surface model. The grid used for China is 1 km × 1 km. For each cell, the type of ecosystem is known: cropland, grassland, woodland and aquatic ecosystem. For each of the four ecosystems, it is possible to calculate the food production (kcal, proteins and fat) per unit area. Based on the need per person of 2298 kcal, 77 g proteins and 67 g of fat, it is possible to calculate the carrying capacity for each cell in the grid. All three needs must be provided per person by the calculation of the carrying capacity. The results are indicated by use of different colors in the grids to facilitate the interpretation.

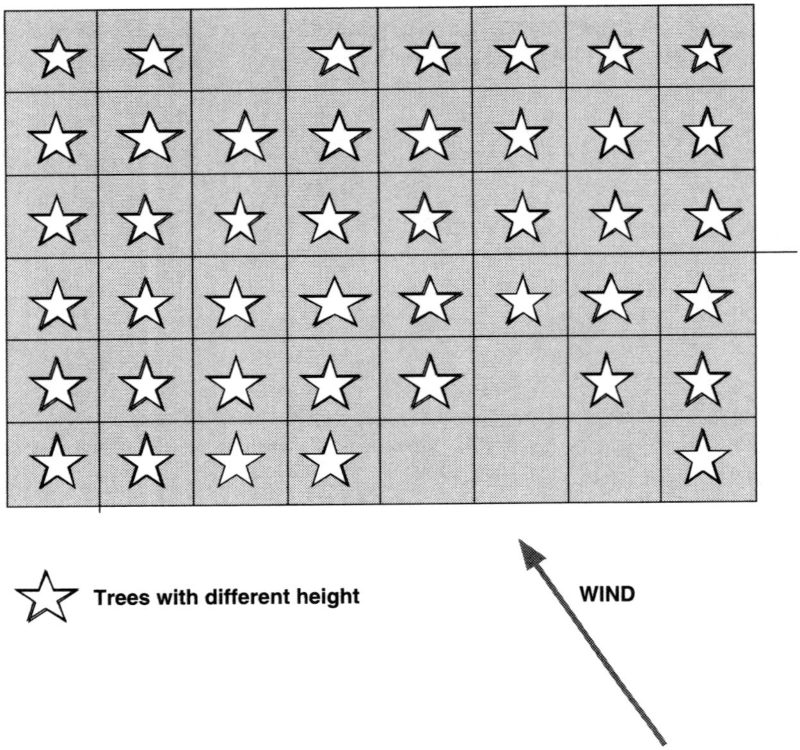

Figure 6.7: The distribution of trees in the forest is shown at time t1. The trees have turned over in four cells according to the rules (see the text).

Malanson *et al.* [8] give another example. They focus on the ecological processes and spatial patterns as a result of deforestation. They consider in their model for each cell of the grid, decomposition of the remnants after deforestation and a time lag before the composition and diversity are changed accordingly. They also consider the recovery and the time lag in the recovery processes. The differences in the different grids due to basic properties are used to calculate the changes in each grid according to the decomposition of the remnants and the recovery processes. The result is different habitat quality indicated as five different levels by use of five different colors. It is, of course, possible to calculate various indices and present the results similarly as the habitat quality.

6.7 Use of IBM in research: determination of the role of conjugation in the evolution

The evolution has applied a wide spectrum of different evolutionary mechanisms to accelerate the evolution [9, 10]. An IBM has been applied to examine the role of one of these mechanisms, namely the conjugation. The model is based on a software developed by Chon *et al.* and is available on a CD enclosed in the book *Handbook of Ecological Modelling and Informatics* by Jørgensen *et al.* [11]. After presentation of the applied IBM, the model is used to illustrate how conjugation is able to increase the biomass density and the eco-exergy of a population. Eco-exergy for a population is the amount of energy that can do work compared with the inorganic

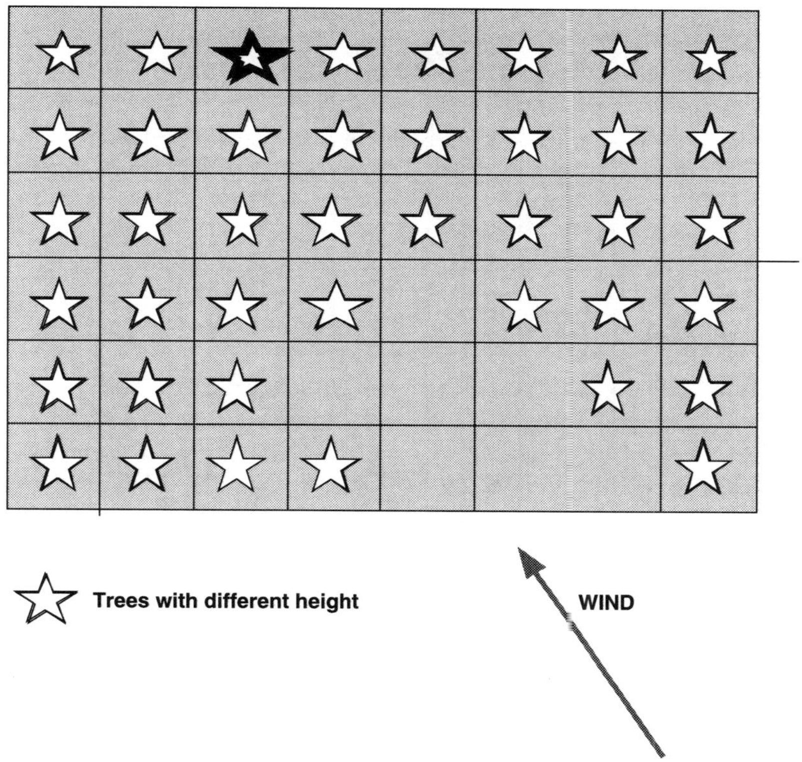

Figure 6.8: The distribution of trees in the forest at time t2. In three cells, the trees have been turned over by wind according to the rules and in one cell, a tree (indicated as a star with bold lines) has been formed by natural seed spreading.

matter contained in the population, that is, at thermodynamic equilibrium – for further details see Chapter 7. It implies that the conjugation is able to move the population system further away from thermodynamic equilibrium [10]. Furthermore, it is examined as to what extent the conjugation is able to increase the biodiversity and Shannon Index. The case demonstrates an application of IBM in research, but it is clear that the model also could have considered a spatial distribution of a population where the individuals have different properties.

6.7.1 The model

The model is developed to illustrate the dynamics of a population to demonstrate survival of individuals with different types of genes under the constraints of food competition and toxic effects.

Individuals carry different gene information determining metabolic efficiency and toxin susceptibility. While some individuals have an advantage in metabolic efficiency, other individuals have better resistance to toxins. Individuals carrying the genes move around on a lattice space to compete for the food with other individuals and with possible exposure to toxins. Health scores are generated as an attribute to each individual and to determine life processes (e.g., reproduction, death). For reproduction, binary fission was assumed, but exchange of genetic information could also occur through conjugation. Health scores are accumulated: positively with consumption of

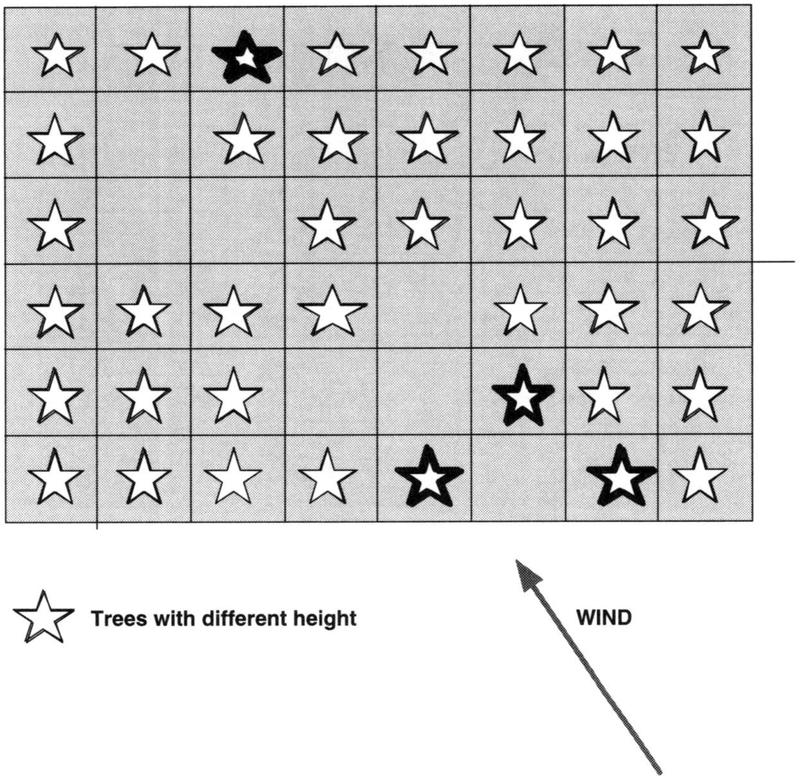

Figure 6.9: The distribution of trees in the forest at time t3. Three more trees have been turned over since time t2 and three trees were formed by natural seed spreading (stars with bold lines). Compare with the Figs 6.6–6.8 and it is understandable that the lattice approach or cellular automata approach is producing a film that is showing the distribution of trees in the forest as function of time. The model has 48 cells and 48 state variables corresponding to the presence and absence of trees in the cells with indication of the height and the forcing function is the wind speed, while the rules may be compared with equations in the biogeochemical models.

nutrients and negatively with exposure to toxins. The phenotypic outcome such as survival by competition would determine the gene levels of the population, while the gene levels control attributes of individuals. In the long run, the population has a higher chance of accumulating the most adaptable genes (i.e., survival of the fittest).

For simulation, an imaginary lattice with 800 times 800 units was assigned in two dimensions. The system was run in 1,000 time steps and the software Visual Basic was used for constructing the IBM. The computer program is provided as an IBM for the population dynamics as a result of conjugations.

6.7.2 State variables

The attributes such as age, health score and location of the individuals (x and y coordinates) were provided as attributes for individuals in the model. Additionally, health scores were used to address life processes of the individuals. Environmental factors, nutrients (food) and toxins were provided in the system. The nutrients in the lattice were consumed accordingly by the entering individuals and were re-supplied regularly to the empty space. If the individuals were exposed to toxins while they move around, toxins were also consumed after exposure with the cost of decrease in the individual's health score. Toxins were also re-supplied to the system on the regular basis to the empty space at the time when the nutrients were re-supplied. At the population level, the total densities and densities of individuals in different types of gene information were used as variables.

6.7.3 Model structure

As stated above, life processes were presented in individuals such as movement, consumption of nutrients, exposure to toxin, reproduction and death. Individuals move according to random walk (one unit per time step in the Neumann type neighbors) if there is no constraint. Each interior site (i, j) (where i = 2…n – 1 and j = 2…n – 1) has eight immediate neighbor lattices (i – 1, j – 1); (i – 1, j); (i – 1, j +1); (i, j–1); (i, j+1); (i+1, j–1); (i+1, j) and (i+1, j+1). If a nutrient is located at one of the neighbor lattices, the individual moves to that lattice. If more than one nutrient exists in the neighborhood area, coin tossing is carried out for determining the selection. Only one unit of nutrient is consumed by one individual at one time step. If the nutrients do not exist in the neighborhood lattices, the individuals move at random. If toxins are located at the neighbor lattices, the individuals are exposed to only one of the toxin randomly.

The two types of genes for metabolic efficiency and toxin susceptibility were generated separately. The genes were assumed to control the degree of metabolic efficiency and toxin susceptibility at some fixed rates (e.g., 0.5, 0.25 and 0.1 for metabolic efficiency), while the maximum gene information was regarded as 0.5. Gene information was converted to phenotypic properties through health scores. The maximum score of metabolic efficiency and toxin susceptibility is assumed to be 20 points. Considering, if the health score previously accumulated is 40 points and the individual has gene information for metabolic efficiency as 0.5, the updated health score will be 50 points (= 40 + (20 (maximum nutrient score)) × 0.5). The health score would decrease similarly when the individual was exposed to toxins (20 (maximum toxin score) toxin susceptibility).

The health scores would be accordingly accumulated with history of food taking and toxin exposure as time progresses. If health score is >50 and age is older than 3 time steps, reproduction will occur, while individuals will die if the health score is < or = 0 or age is more than 300 time steps. Individuals reproduce by fission, but conjugation could occur on a stochastic basis in case other individuals are located at the neighbor lattices. If two or more individuals are located at the neighbor sites, coin tossing is carried out for determination of the counterpart of the conjugation. Only one type of genes is selected for exchange at random. Suppose that an individual (say A) has 0.5 for metabolic efficiency and 0.5 for toxin susceptibility as gene information, and the individual A conjugates with another individual (say C) with 0.1 metabolic efficiency and 0.05 toxin susceptibility. If the gene for metabolic efficiency is selected for conjugation after coin tossing, the individual A will have 0.1 for metabolic efficiency and 0.5 for toxin susceptibility, while the individual C will carry 0.5 for metabolic efficiency and 0.05 for toxin susceptibility. For simplicity of modelling, we assume that conjugation occurs at 100% when the individual at the center lattice was placed closed to the neighbor individuals.

6.7.4 Parameters and forcing functions

The variables presenting individual (e.g., health score, location, age) and population properties (e.g., density in different types) were updated and provided to the system in each time step.

In addition, nutrients and toxins were provided as environmental factors in the system. Only one, either nutrient or toxin, is allowed per lattice. The parameters for the supply of nutrients (initially 20% of the total lattice and re-supply in 10% of the empty space in each 100 time step), and toxins (initially 20% of the total lattice and re-supply in extra 1% of the empty space in each 100 time steps) are also incorporated to the system. For operation of the model, initial conditions for input variables were provided: age (0 time step, (range; 0–300)) and health score (40 points, (range; 0–100 points)). Three types of individuals (10 of each type) with different gene information were assigned as initial population, for instance, as listed below:

- Type A: metabolic efficiency - 0.5, toxin susceptibility - 0.5;
- Type B: metabolic efficiency - 0.25, toxin susceptibility - 0.25;
- Type C: metabolic efficiency - 0.1, toxin susceptibility - 0.05.

Type A is strong in metabolic efficiency (0.5) and weak in toxin resistance (high in toxin susceptibility (0.5), while Type C is the opposite of Type A. Type B is in the intermediate range in metabolic efficiency (0.25) and toxin susceptibility (0.25). Each individual was placed at random in the 2D lattice space initially.

6.7.5 Model results

Through conjugation, nine types were produced in combination of two types of genes:

- A-A: metabolic efficiency – 0.5, toxin susceptibility – 0.5;
- A-B: metabolic efficiency – 0.5, toxin susceptibility – 0.25;
- A-C: metabolic efficiency – 0.5, toxin susceptibility – 0.05;
- B-A: metabolic efficiency – 0.25, toxin susceptibility – 0.5;
- B-B: metabolic efficiency – 0.25, toxin susceptibility – 0.25;
- B-C: metabolic efficiency – 0.25, toxin susceptibility – 0.05;
- C-A: metabolic efficiency – 0.1, toxin susceptibility – 0.5;
- C-B: metabolic efficiency – 0.1, toxin susceptibility – 0.25;
- C-C: metabolic efficiency – 0.1, toxin susceptibility – 0.05.

Figure 6.10 shows the population dynamics obtained by the simulations. Dominance types selectively appeared as the time progressed. Without conjugation (Fig. 6.10a), no recombination of genes occurred and the three original types (A-A, B-B and C-C) were only produced. In this case, Type A-A that carries genes for maximum rate of metabolism (0.5) and the minimum rate of toxin resistance (i.e., the highest level of toxin susceptibility (0.5)) was dominant the first more than 300 time steps compared with the other types (Fig. 6.10a). At some lower levels, Type B-B was also established. The other Type (C-C), however, was not established by the simulation. The simulation results revealed that the system constrained by food competition and toxin exposure could serve as a selection force for determining dominance of the traits in ecotoxicology.

The organisms with higher efficiency in food consumption appeared to be advantageous since nutrients were supplied at a relatively higher level, initially. Densities of the A-A and B-B types, however, did not persist for a long time. They temporarily dominated and declined soon. In the middle of simulation, the total density remained at low levels for a substantial period. This is due to the depletion of food at this stage. Consequently, population could not recover readily after the nutrients were exploited by the organisms. However, populations started to recover again later around 600 time steps. Extra nutrients are accumulated during the periods of low densities after

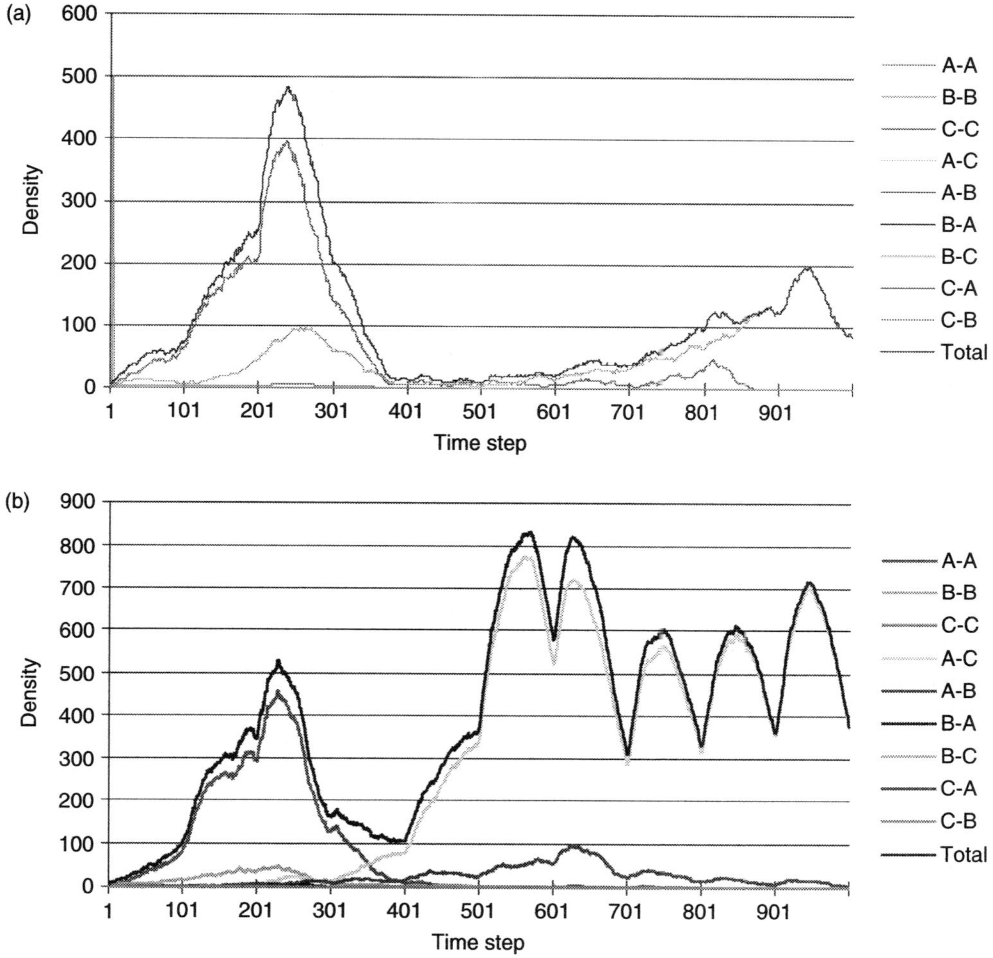

Figure 6.10: Results of the described test simulations: (a) is without conjugation; (b) with 100% conjugation. The evolutionary advantages of conjugations are seen very clearly from the results. After 1000 time steps, the total density representing biomass and eco-exergy is about three times higher with conjugation than without conjugation, and AC is completely dominant (almost equal to total) after time step 400.

the dominant type had been declined in the early period. The same types, Type A-A and Type B-B, became dominant again, ensuring that Type C-C was not adapted to the system as well as the other two types. However, both types were not recovered strongly and started to decline again around 1,000 time steps. The gene of Type C-C is clearly not competitive. The genotype–phenotype relationships are, however, complicated and further study is required regarding the gene–environment causality relationships.

When conjugation (100%) was allowed for gene exchange, however, population dynamics appeared differently (Fig. 6.10b). Initially, Type A-A was dominant similar to the case without conjugation, but Type A-C gradually became dominant to exceed Type A-A eventually. Type A-C was obtained from conjugation and continuously increased as the time progressed. This is understandable: Type A-C carries the most adaptive gene information (maximum for both metabolic efficiency and toxin resistance). At the later period of simulation, Type A-C was completely dominant, prevailing over the entire lattice space. The population also showed periodic changes. This may be due to coupling with the periodic nutrient re-supply, as stated before.

The model results show that IBMs are favorably applied when addressing complex phenomena in ecotoxicology. Comprehensive understanding was available on addressing the properties residing in the complex population dynamics under influence of food and a toxic substance with and without conjugations. Interaction with other individuals (i.e., competition) and environmental factors was demonstrated in IBMs, being efficiently fabricated with the model structure interlinking the upper and lower levels in the hierarchy of biological systems. The phenotypic traits were sensitively dynamic to the constraints of the genotypes. However, this study only proposed an initial stepstone in IBMs toward the understanding the role of conjugations in the evolution. Further study is required in the future on many different aspects regarding computational analysis on system dynamics residing in the individual-population relationships, quantitative characterization of the emergent properties produced from the model, biological/ecological experiments linking genes-behavior-population relationships and technical development in IBM construction.

6.7.6 Discussion and conclusions

According to the obtained results conjugation seems to give the evolution a significant push forward toward higher eco-exergy (the distance from the thermodynamic equilibrium – for further details see next chapter) and more possibilities for moving further away from thermodynamic equilibrium. Higher species richness implies more possibilities for further genetic improvements by mutations and sexual re-combinations and more possibilities for establishment of larger ecological networks. Similarly, sexual recombination affords a vast scope for change and for new evolutionary possibilities, yet the genetic integrity is still maintained. The diversity that, for instance, 1000 individual of a mammal species with 25,000 genes and totally 6 billion bits of information is astronomic. The introduction of conjugation and later, sexual recombination in the evolution have given an enormous rise in the variability that has been of utmost significance for all the evolutionary processes.

References

[1] Hu, W., Jørgensen, S.E. & Zhang, F., A vertical-compressed three-dimensional ecological model in Lake Taihu. *Ecological Modelling*, **190**, pp. 367–398, 2006.

[2] Mogensen, B. & Jørgensen, S.E., Modelling the distribution of chromium in a Danish fjord. *Proceedings of Ist International Conference on State of the An in Ecological Modelling*, Copenhagen, 1978, ed. S.E. Jørgensen, International Society for Ecological Modelling: Copenhagen, pp. 367–377, 1979.

[3] Weber, W.J., *Physicochemical Processes for Water Quality Control*, Wiley Interscience: New York, 1972.

[4] Jørgensen, S.E., Modelling the distribution and effect of heavy metals. *Ecological Modelling*, **6**, pp. 199–222, 1979.

[5] Jørgensen, L.A., Jørgensen, S.E. & Nielsen, S.N., *Ecotox*, CD, 2000.

[6] Metz, J.A. & Dickmann, O., *The Dynamics of Physiologically Structured Populations*, Springer Verlag: Berlin, 1986.

[7] Yue, T.X., Tian, Y.Z., Liu, J.Y. & Fan, Z.M., Surface modelling of human carrying capacity of terrestrial ecosystems in China. *Ecological Modelling*, **214**, pp. 168–180, 2008.

[8] Malanson, G.P., Wang, Q. & Kupfer, J.A., Ecological processes and spatial patterns before, during and simulated deforestation. *Ecological Modelling*, **202**, pp. 397–409, 2007.

[9] Jørgensen, S.E., *Evolutionary Essays: A Thermodynamic Interpretation of the Evolution*, Elsevier: Amsterdam, 2008.

[10] Jørgensen, S.E. & Fath, B., *A New Ecology. A Systems Approach*, Elsevier: Amsterdam, Oxford, 2007.

[11] Jørgensen, S.E., Chon, T. & Recknagel, F., *Handbook of Ecological Modelling and Informatics*, WIT Press: Southampton, 2009.

Exercises/problems – Chapter 6

1. Under which circumstances can the distance be used as an independent variable? Give at least two examples where it seems fully acceptable to eliminate time as the independent variable and replace time by distance.
2. Why could the chromium model presented in Section 6.3 not be applied directly for the distribution of tin and mercury? Mention at least three additional state variables that should be included in a model of the mercury or tin distribution in a fjord.
3. Which factors are decisive for the number of boxes that would be needed in a river model with many different discharges of wastewater and drainage water?

CHAPTER 7

Structurally dynamic models (SDMs)

7.1 Basic concepts of structurally dynamic modelling

Models of ecosystems attempt to capture the characteristics of ecosystems. However, ecosystems differ from most other systems by being extremely adaptive, having the ability of self-organization and having a large number of feedback mechanisms. The real challenge to ecological modelling is therefore: How can we construct models that are able to reflect these characteristics? Section 7.2 will focus on the characteristic of ecosystems, which we should consider in our modelling approach. Section 7.3 will present the thermodynamic variable eco-exergy or work capacity and show how it can be calculated for ecological models. Eco-exergy is used to translate Darwin's "Survival of Fittest" into thermodynamic and can therefore be used as a goal function in models that describe the changes to better fitted properties for survival of the biological model components under the steadily changed prevailing conditions.

Section 7.4 is devoted to how to develop what is named structurally dynamic models (SDMs) or variable parameter models. The section will present the advantages and disadvantages of this model type and where it is most recommendable to consider to apply SDM. Sections 7.5–7.8 present four illustrative examples of SDMs. All the illustrations will be explained in details.

This model type will most probably be used increasingly in the coming years in our endeavor to make better prognoses, because reliable prognoses can only be made by models with a correct description of ecosystem properties including the ability to change the structure and the properties of key species. If our models do not describe properly adaptation and possible shifts in species composition, the prognoses will inevitably be more or less incorrect, because these two properties may be very important for the development of ecosystems and their reactions to changing forcing functions.

SDMs can be developed by the use of expert knowledge. Artificial intelligence systems can be applied to translate the expert knowledge into rules that can guide the model changes that accounts for adaptation and shifts in species compositions.

It is also possible to apply a goal function to develop SDMs. The thermodynamic variable, eco-exergy or work capacity can as mentioned above be applied as a goal function. The presentation of this concept in details will be given in Section 7.3, as indicated above. There is not yet any software available that can be used for a general development of SDM based on the use of a goal function. It is, however, possible to develop for instance a C++ program for the development of SDMs. Downloadable SDMs of eutrophication for both deep and shallow lakes and wetlands have been developed by the author of this book for UNEP under the name Pamolare. They can be found on the home page www.unep.org.jp.

Notice that Pamolare offers two SDMs of lakes, for a deep lake with thermocline under the name Pamolare 1 together with three other lake eutrophication models and for a shallow lake under the name Pamolare 2. It is also possible to apply STELLA to develop SDMs, but it is more time consuming than application of ready-to-go software of SDMs. Both the use of ready-to-go software and STELLA will be considered later in the chapter and concrete illustrative examples will be given for both the methods.

7.2 Ecosystem characteristics

Ecology deals with irreducible systems [1–9]. We cannot design simple experiments which will reveal a relationship that can in all detail be transferred from one ecological situation in one ecosystem to another situation in another ecosystem. That is possible, for instance, with Newton's laws of gravity, because the relationship between forces and acceleration is reducible. The relationship between force and acceleration is linear, but growth of living organisms is dependent on many interacting factors, which again are functions of time. Feedback mechanisms, self-organization and autocatalysis will simultaneously regulate all the factors and rates, and they also interact, and are also functions of time, as pointed out by Straskraba [10, 11].

Table 7.1 shows the hierarchy of regulation mechanisms that are operating at the same time. From the discussion above, the complexity alone clearly prohibits the reduction to simple relationships that can be used repeatedly.

An ecosystem has so many interacting components that it is impossible to examine all these relationships and even if we could, it would not be possible to separate one relationship and examine it carefully to reveal its details, because the relationship is different when it works in nature with interactions from the many other processes, from when we examine it in a laboratory with the relationship completely separated from the other ecosystem components.

A known phrase in system ecology is: “everything is linked to everything” or: “the whole is greater than the sum of the parts”. It implies that it may be possible to examine the parts by reduction to simple relationships, but when the parts are put together they will form a whole, which behaves differently from the sum of the parts. This statement requires a more detailed discussion of how an ecosystem works.

Table 7.1: The hierarchy of regulating feedback mechanisms [11].

Level	Explanation of regulation process	Exemplified by phytoplankton growth
1	Rate by concentration in medium	Uptake of phosphorus in accordance with phosphorus concentration
2	Rate by needs	Uptake of phosphorus in accordance with intracellular concentration
3	Rate by other external factors	Chlorophyll concentration in accordance with previous solar radiation
4	Adaptation of properties	Change of optimal temperature for growth
5	Selection of other species	Shift to better fitted species
6	Selection of other food web	Shift to better fitted food web
7	Mutations, new sexual recombinations and other shifts of genes	Emergence of new species or shifts of species properties

Allen and Starr [12] claim that the latter statement is correct, because of the evolutionary potential that is hidden within living systems. The ecosystem contains within itself the possibilities of becoming something different, that is, of adapting and evolving. The evolutionary potential is linked to existence of microscopic freedom, represented by stochasticity and non-average behavior, resulting from the diversity, complexity and variability of its elements.

Underlying the taxonomic classification is the microscopic diversity, which only adds to the complexity to such an extent that it will be completely impossible to cover all the possibilities and details of the observed phenomena. We attempt to capture at least a part of the reality by use of models. A model seems to be the only useful tool when we are dealing with irreducible systems. However, we need many models simultaneously to capture a more complete image of reality. It seems to be our only possibility to deal with the very complex living systems.

This has been acknowledged by the holistic ecology or systems ecology, while the more reductionistic ecology attempts to understand ecological reactions by analysis of one or at the most a few processes, which are related to one or two components. Both analyses and syntheses are needed in ecology and the analysis is a necessary foundation for the synthesis, but it may lead to wrong scientific conclusions to stop at the analysis. Analysis of several interacting processes may give a right result of the processes under the analyzed conditions, but the conditions in ecosystems are constantly changing and even if the processes were unchanged (which they very rarely are), it is not possible to overview the analytical results of so many simultaneously working processes. Our brain simply cannot overview what will happen in a system where, for instance, only six interacting processes are working simultaneously.

The reductionistic approach does not consider:

1. The basic conditions determined by the external factors for our analysis are constantly changing in the real world and the analytical results, where only one or a few factors vary, are therefore not necessarily valid in the system context.
2. The interaction from all the other processes and components may change the processes and the properties of all biological components significantly in the real ecosystem and the analytical results are therefore not valid at all.
3. A direct overview of the many processes simultaneously working is not possible and wrong conclusions may be the result, if it is attempted.

The conclusion is, therefore, that we need a tool to overview and synthesize the many interacting processes. The synthesis may in the first instance just be "putting together" the various analytical results, but afterward we need to make changes to account for an additional effect, resulting from the fact, that the processes are working together and thereby become more than the sum of the parts – they show in other words, a synergistic effect – a symbiosis. By a network analysis, it is possible to show how important the indirect effects are compared to the direct effects in an ecological network.

It is clear from this discussion that the complexity of ecosystems has set the limitations for our understanding and for the possibilities of proper management. We cannot capture the complexity as such with all its details, but we can understand how ecosystems are complex and we can set up a realistic strategy for how to get sufficient knowledge about the system – not knowing all the details, but still understanding and knowing the behavior and the important reactions of the system. It means that we can only try to reveal the basic properties behind the complexity.

We have no other choice than to go holistic. The results from the more reductionistic ecology are essential in our effort "to go to the root" of the system properties of ecosystems, but we need systems ecology, which consists of many new ideas, approaches and concepts, to follow the path to the roots of the basic system properties of ecosystems. The idea may also be expressed in another way: we cannot find the properties of ecosystems by analyzing all the details, because they are simply too many, but only by trying to reveal the system properties of ecosystems by examination of the entire systems.

The number of feedbacks and regulations is extremely high and makes it possible for the living organisms and populations to survive and reproduce inspite of changes in external conditions. These regulations correspond to levels 3 and 4 in Table 7.1. Numerous examples can be found in the literature. If the actual properties of the species are changed the regulation is named adaptation.

Phytoplankton is, for instance, able to regulate its chlorophyll concentration according to the solar radiation. If more chlorophyll is needed because the radiation is insufficient to guarantee growth, more chlorophyll is produced by the phytoplankton. The digestion efficiency of the food for many animals depends on the abundance of the food. The same species may be of different sizes in different environments, depending on what is most beneficial for survival and growth. If nutrients are scarce, phytoplankton becomes smaller and vice versa. In this latter case, the change in size is a result of a selection process, which is made possible because of the distribution in size.

The feedbacks are also constantly changing, that is, the adaptation itself is adaptable in the sense that if a regulation is not sufficient, another regulation process higher in the hierarchy of feedbacks (see Table 7.1) will take over. The change in size within the same species is, for instance, only limited. When this limitation has been reached, other species may take over (see level 5 and 6 in Table 7.1). It implies that not only the processes and the components, but also the feedbacks can be replaced, if it is needed to achieve a better utilization of the available resources.

An ecosystem is a very dynamic system. All its components and particularly the biological ones are steadily moving and their properties are steadily modified, which is why an ecosystem will never return to the same situation again.

Every point is different from any other point and therefore offering different conditions for the various life forms.

This enormous heterogeneity explains why there are so many species on earth. There is, so to say, an ecological niche for "everyone" and "everyone" may be able to find a niche where he is best fitted to utilize the resources.

Margalef [13, 14] claims that ecosystems are anisotropic, meaning that they exhibit properties with different values, when measured along axes in different directions. It means that the ecosystem is not homogeneous in relation to properties concerning matter, energy and information, and that the entire dynamics of the ecosystem works toward increasing the differences.

These variations in time and space make it particularly difficult to model ecosystems and to capture the essential features of ecosystems.

Ecosystems and their biological components, the species, evolve steadily and in the long term perspective toward higher complexity.

Darwin's theory describes the competition among species and states that the species that are best fitted to the prevailing conditions in the ecosystem will survive. Darwin's theory can, in other words, describe the changes in ecological structure and the species composition, but cannot directly be applied quantitatively, for example, in ecological modelling (see, however the next section).

All species in an ecosystem are confronted with the challenge: how is it possible to survive or even grow under the prevailing conditions? The prevailing conditions are considered as factors influencing the species, that is, all external and internal factors including those originating from other species. This explains the coevolution, as any change in the properties of one species will influence the evolution of the other species.

All natural external and internal factors of ecosystems are dynamic – the conditions are steadily changing and there are always many species waiting in the wings, ready to take over, if they are better fitted to the emerging conditions than the species dominating under the present conditions. There is a wide spectrum of species representing different combinations of properties available for the ecosystem. The question is, which of these species are best able to survive and grow under the present conditions and which species are best able to survive and grow under the conditions one time step further and two time steps further and so on? The necessity in Monod sense is given by the prevailing conditions – the species must have genes or may be rather phenotypes (meaning properties) that match these conditions, to be able to survive. But the natural external factors and the genetic pool available for the test may change randomly or by "chance".

Steadily, new mutations (misprints are produced accidentally) and sexual recombinations (the genes are mixed and shuffled) emerge and give steadily new material to be tested by the question: which species are best fitted under the conditions prevailing just now?

These ideas are illustrated in Fig. 7.1. The external factors are steadily changed and some even relatively fast – partly at random, such as the meteorological or climatic factors. The species of the system are selected among the species available and represented by the genetic pool, which again is slowly, but surely changed randomly or by "chance". The selection in the Fig. 7.1 includes the level 4 of Table 7.1. It is a selection of the organisms, which possess the properties best fitted to the prevailing organisms according to the frequency distribution. What is named ecological development is the changes over time in nature caused by the dynamics of the external factors, giving the system sufficient time for the reactions.

Evolution, on the other hand, is related to the genetic pool It is the result of the relation between the dynamics of the external factors and the dynamics of the genetic pool. The external factors steadily change the conditions for survival and the genetic pool steadily comes up with new solutions to the problem of survival.

The species are continuously tested against the prevailing conditions (external as well as internal factors) and the better they are fitted, the better they are able to maintain and even increase their biomass. The specific rate of population growth may even be used as a measure for the fitness [15]. But the property of fitness must, of course, be inheritable to have any effect on the species composition and the ecological structure of the ecosystem in the long run.

Natural selection has been criticized for being a tautology: fitness is measured by survival and survival of the fittest therefore means survival of the survivors. However, the entire Darwinian theory including the above mentioned three assumptions cannot be conceived as a tautology, but may be interpreted as follows: the species offer different solutions to survival under given prevailing conditions and the species that have the best combinations of properties to match the conditions have also the highest probability of survival and growth.

The evolution moves toward increasing complexity in the long run. The fossil records have shown a steady increase of species diversity. There may be destructive forces – for instance man-made pollution or natural catastrophes – for a shorter time, but the probability that

1. new and better genes are developed; and
2. new ecological niches are utilized

will increase with time. The probability will even – again excluding the short time perspective – increase faster and faster, as the probability is roughly proportional to the amount of genetic material on which the mutations and new sexual recombinations can be developed.

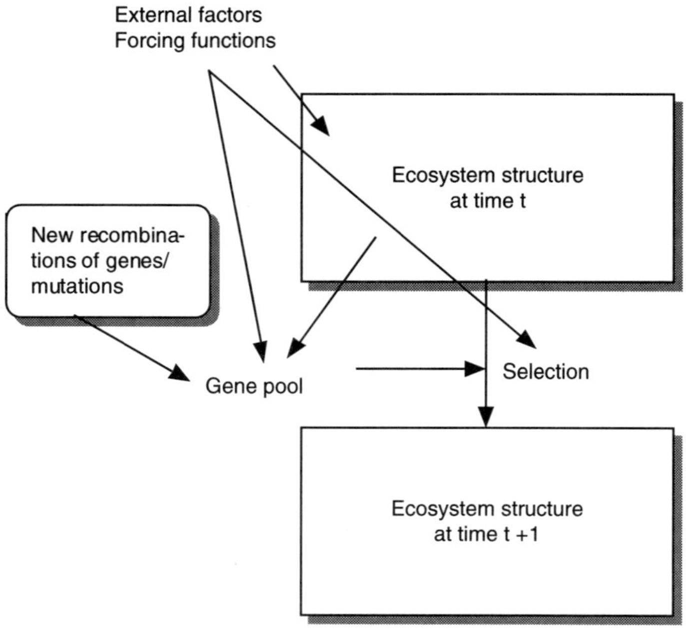

Figure 7.1: Conceptualization of how the external factors steadily change the species composition. The possible shifts in species composition are determined by the gene pool, which is steadily changed due to mutations and new sexual recombinations of genes. The development is, however, more complex. This is indicated (i) by arrows from "external factors" to "structure" and "selection" and (ii) an arrow from "structure" to "gene pool" to account for the possibilities that the species can to a certain extent change their own gene pool. In addition, it is also possible for the organisms to modify their environment and thereby the forcing functions of the system. Several mechanisms for this possibility can be found in the literature.

Jørgensen and Fath [16] presented a set of tentative basic propositions for ecosystems in 2004. One of the propositions based on eco-exergy will be presented in the next section.

Recently, June 2005, at a brainstorming meeting about the possibilities to apply a wide general ecosystem theory, it was concluded that the basic propositions presented by Jørgensen and Fath approximately cover what a basic ecosystem theory should contain, and the best method to present an ecosystem theory for ecologists in general was to use the basic properties of ecosystem, which in principle only is another more down-to-earth presentation of an ecosystem theory. The properties that explain the dynamic and the reactions of ecosystems are in accordance to the book resulting from the brainstorming meeting – *A New Ecology, A Systems Approach*, edited by Jørgensen and Fath [17]:

1. Ecosystems are physical and ontic open, meaning that they can exchange mass, energy and information with the surroundings and that it is not possible to make exact predictions on their development due to their enormous complexity; see the discussion above.
2. Ecosystems have directionality.

3. Ecosystems have connectivity.
4. Ecosystems have emergent hierarchies.
5. Ecosystems have a complex dynamics (growth but also reactions toward disturbances).

7.3 The thermodynamic concept of eco-exergy

Exergy is defined as the work capacity the system can perform when brought into thermodynamic equilibrium with the environment, considering the difference in temperature (heat energy), the difference in pressure (pressure energy), the difference in altitude (mechanical potential energy), the difference in voltage (electrical energy) and the difference in chemical potential (chemical energy) to mention the most applied energy forms.

This form of exergy we will name here as technological exergy. Technological exergy is not practical to use in the ecosystem context, because it presumes that the environment is the reference state which means for an ecosystem the next ecosystem. As the energy embodied in the organic components and the biological structure and information contributes far most to the exergy content of an ecosystem, there seems, furthermore, no reason to assume a (minor) temperature and pressure difference between the ecosystem and the reference environment. Eco-exergy is defined (see Fig. 7.2) as the work the ecosystem can perform relative to the same ecosystem at the same temperature and pressure but at thermodynamic equilibrium, where there are no gradients and all components are inorganic at the highest possible oxidation state. Under these circumstances we can calculate the exergy, which has been denoted eco-exergy to distinguish from the technological exergy, as coming entirely from the chemical energy:

$$\sum_{c}(\mu_c - \mu_{co})N_i.$$

Eco-exergy has successfully been used to develop SDMs [9], as a holistic ecological indicator [17] and as an important variable to describe ecosystem dynamics.

Eco-exergy represents the non-flow biochemical exergy. It is determined by the difference in chemical potential ($\mu_c - \mu_{co}$) between the ecosystem and the same system at thermodynamic equilibrium. This difference is determined by the activities (approximated by the use of the concentrations) of the considered components in the system and in the reference state (thermodynamic equilibrium), as it is the case for all chemical processes. We can measure the concentrations in the ecosystem, but the concentrations in the reference state (thermodynamic equilibrium) could be based on the usual use of chemical equilibrium constants. Eco-exergy is a concept close to Gibbs free energy, but opposite to Gibbs free energy, eco-exergy has a different reference state from case to case (from ecosystem to ecosystem) and it can also be used far from thermodynamic equilibrium, while Gibbs free energy in accordance to its exact thermodynamic definition is a state function close to thermodynamic equilibrium. In addition, eco-exergy of organisms is mainly embodied in the information content and should, therefore not to be considered the same as the chemical energy of fossil fuel.

As ($\mu_c - \mu_{co}$) can be found from the definition of the chemical potential replacing activities by concentrations, we get the following expressions for the exergy:

$$Ex = RT\sum_{i=0}^{i=n} C_i \ln C_i / C_{i,o} \tag{7.1}$$

where R is the gas constant (8.317 J/ K moles = 0.08207 liter. atm/K moles), T is the temperature of the environment, while C_i is the concentration of the i'th component expressed in a suitable

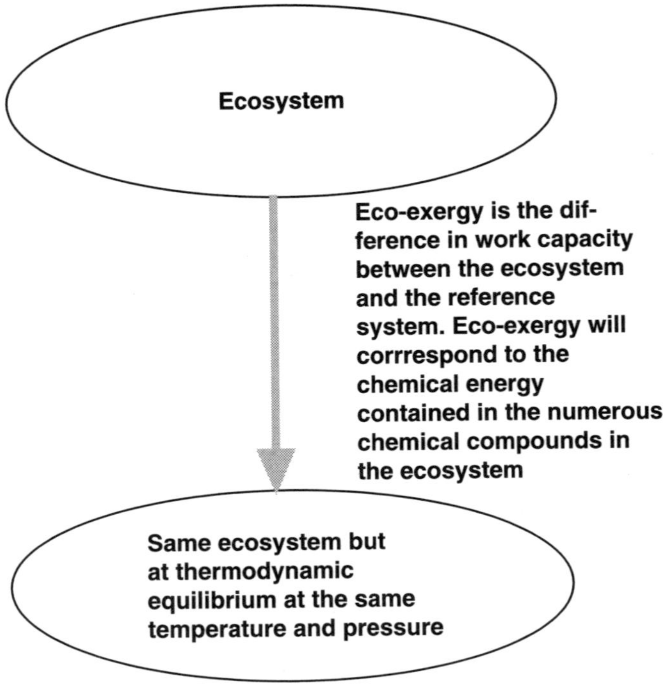

Figure 7.2: The definition of eco-exergy is shown. The work capacity in the ecosystem in form of the chemical energy of the many different and complex chemical compounds relative to the reference system is the eco-exergy. The reference system is the same ecosystem but at thermodynamic equilibrium, that is, a homogeneous system without life. All the chemical compounds are inorganic and there are no gradients.

unit. $C_{i,o}$ is the concentration of the i'th component at thermodynamic equilibrium and n is the number of components. $C_{i,o}$ is, of course, a very small concentration (except for i = 0, which is considered to cover the inorganic compounds), corresponding to a very low probability of forming complex organic compounds spontaneously in an inorganic soup at thermodynamic equilibrium. $C_{i,o}$ is even lower for the various organisms, because the probability of forming the organisms is very low with their embodied information which implies that the genetic code should be correct.

By using this particular exergy based on the same system at thermodynamic equilibrium as reference, the eco-exergy becomes dependent only on the chemical potential of the numerous biochemical components.

It is possible to distinguish, in eqn (7.1), between the contribution to the eco-exergy from the information and from the biomass. We define p_i as c_i/A, where

$$A = \sum_{i=1}^{n} c_i$$

is the total amount of matter density in the system. With introduction of this new variable we get:

$$Ex = ART\sum_{i=1}^{n} p_i \ln p_i/p_{io} + A \ln A/A_o \tag{7.2}$$

As $A \approx A_o$, eco-exergy becomes a product of the total biomass A (multiplied by RT) and Kullback's measure of information:

$$K = \sum_{i=1}^{n} p_i \ln(p_i/p_{io}) \tag{7.3}$$

where p_i and p_{io} are probability distributions, a posteriori and a priori to an observation of the molecular detail of the system. It means that K expresses the amount of information that is gained as a result of the observations. For different organisms that contribute to the eco-exergy of the ecosystem, the eco-exergy density becomes cRTln (p_i/p_{io}), where c is the concentration of the considered organism. RTln (p_i/p_{io}), denoted as β is found by calculation of the probability to form the considered organism at thermodynamic equilibrium, which would require that organic matter is formed and that the proteins (enzymes) controlling the life processes in the considered organism have the right amino acid sequence. These calculations can be seen in the book by Jørgensen and Svirezhev [18]. In the latter reference, the latest information about the β-values for various organisms is presented (see Table 7.2). For human the β-value is 2173, when the eco-exergy is expressed in detritus equivalent or 18.7 times as much or 40635 kJ/g, if the eco-exergy should be expressed as kJ and the concentration unit g/unit of volume or area. The β-value has not surprisingly increased as a result of the evolution. To mention a few β-values from Table 7.2: bacteria 8.5, protozoa 39, flatworms 120, ants 167, crustaceans 232, molluscs 310, fish 499, reptiles 833, birds 980 and mammals 2127. The evolution has in other words resulted in a more and more effective transfer of what we could call the classical work capacity to the work capacity of the information. A β-value of 2.0 means that the eco-exergy embodied in the organic matter and the information are equal. As the β-values; seen above are much bigger than 2.0 (except for virus, where the β-value is 1.01 – slightly more than 1.0) the information eco-exergy is the most significant part of the eco-exergy of organisms.

In accordance to eqns (7.2) and (7.3) and the above presented interpretation of these equations, it is now possible to find the eco-exergy density for a model as:

$$\text{Eco-exergy density} = \sum_{i=1}^{i=n} \beta_i c_i \tag{7.4}$$

The eco-exergy due to the "fuel" value of organic matter (chemical energy) is about 18.7 kJ/g (compared with coal: about 30 kJ/g and crude oil: 42 kJ/ g). It can be transferred to other energy forms for instance mechanical work directly, and be measured by bomb calorimetry, which however requires destruction of the sample (organism). The information eco-exergy = $(\beta_i - 1) \times$ biomass or density of information eco-exergy = $(\beta - 1) \times$ concentration. The information eco-exergy takes care of the control and function of the many biochemical processes. The ability of the living system to do work is contingent upon its functioning as a living dissipative system. Without the information eco-exergy, the organic matter could only be used as fuel similar to fossil fuel. But due to the information eco-exergy, organisms are able to make a network of the sophisticated biochemical processes that characterize life. The eco-exergy (of which the major part is embodied in the information) is a measure of the organization [18]. This is the intimate relationship between energy and organization that Schrødinger (1944) in his famous book "What is Life" was struggling to find.

The eco-exergy is a result of the evolution and of what Elsasser (1975) calls re-creativity to emphasize that the information is copied and copied again and again in a long chain of copies where only minor changes are introduced for each new copy. The energy required for the copying process is very small, but it has, of course, required a lot of energy to come to the "mother" copy through the evolution, for instance, from prokaryotes to human cells. To cite Margalef [14] in this context: the evolution provides for cheap – unfortunately often "erroneous" that is, not exact – copies of messages or pieces of information. The information concerns the degree of uniqueness of entities that exhibit one characteristic complexion that may be described.

7.4 Structurally dynamic models

If we follow the modelling procedure proposed in Chapter 1, we will attain a model that describes the processes in the focal ecosystem, but the parameters will represent the properties of the state variables as they are in the ecosystem during the examination period. They are not necessarily valid for another period, because we know that an ecosystem can regulate, modify and change them, if needed as response to the change in the prevailing conditions, determined by the forcing functions and the interrelations between the state variables. Our present models have rigid structures and a fixed set of parameters, reflecting that no changes or replacements of the components are possible. We need, however, to introduce parameters (properties) that can change according to changing forcing functions and general conditions for the state variables (components) to optimize continuously the ability of the system to move away from thermodynamic equilibrium. We will try in this section to answer the question: how can we construct models that are able to reflect the characteristics of ecosystems, particularly the adaptability and the possibilities to change the composition of species corresponding to the levels 4 – 6 in Table 7.1 and as illustrated in Fig. 7.1. Models that account for the change in species composition as well as for the ability of the species – that is, the biological components of our models – to change their properties – that is, to adapt to the prevailing conditions imposed on the species – are called SDMs, to indicate that they are able to capture structural changes. It could be argued that the ability of ecosystems to replace present species with other (level 6 in Table 7.1), better fitted species could be considered by construction of models that encompass all actual species for the entire period that the model attempts to cover. This approach has, however, two essential disadvantages. The model becomes first of all very complex, as it will contain many state variables for each trophic level. It implies that the model will contain many more parameters that have to be calibrated and validated. It will inevitably introduce a high uncertainty to the model and will render the application of the model very case specific [19, 20]. In addition, the model will still be rigid and not give the model the property of the ecosystems of having continuously changing parameters even without changing the species composition [21].

SDMs can be constructed by two different methods as previously mentioned: either by expert knowledge or by introduction of a goal function. If it is known how the properties of the species will change when the prevailing conditions are changed, it is of course possible to introduce this knowledge into the model, which is possible either by formulation of rules or by artificial intelligence. Rules may be exemplified by this example: when the phosphorus concentration is in the range x–y, then the growth rate of phytoplankton will be changing gradually from z to v. For examples of this type of SDMs, see Patten [22]. Artificial intelligence is able to find the rules from interpretation by a computer of a suitable set of observations.

Several goal functions have been proposed. Most models that account for change in species composition or for the ability of the species to change their properties within some limits, apply eco-exergy as goal function.

Bossel [23] uses what he calls six basic orientors or requirements to develop a system model, which can describe the system performance properly. The six orientors are:

1. *Existence:* The system environment must not exhibit any conditions, which may move the state variables out of its safe range.

2. *Efficiency:* The exergy gained from the environment should exceed over time the exergy expenditure.
3. *Freedom of action:* The system reacts to the inputs (forcing functions) with a certain variability.
4. *Security:* The system has to cope with the different threats to its security requirement with appropriate but different measures. These measures either aim at internal changes in the system itself or at particular changes in the forcing functions (external environment).
5. *Adaptability:* If a system cannot escape the threatening influences of its environment, the one remaining possibility is in changing the system itself to cope better with the environmental impacts.
6. *Consideration of other systems:* A system must respond to the behavior of other systems. The fact that these other systems may be of importance to a particular system may have to be considered with this requirement.

Bossel [23] applies maximization of a benefit or satisfaction index based on balancing weighted surplus orientor satisfactions on a common satisfaction scale. The approach is used to select the model structure of continuous dynamic systems and is able to account for the ecological structural properties as presented in Fig. 7.1. The approach seems very promising, but has unfortunately not been applied to ecological systems except in three cases.

Straskraba [11] uses a maximization of biomass as the governing principle. The model computes the biomass and adjusts one or more selected parameters to achieve the maximum biomass at every instance. The model has a routine which computes the biomass for all possible combinations of parameters within a given realistic range. The combination that gives the maximum biomass is selected for the next time step and so on.

Eco-exergy has successfully been used to develop SDMs [9] and as a holistic ecological indicator [6]. The eco-exergy used as a goal function is found by use of an eqn (7.4), while the β-value is found by use of Table 7.2.

The application is based on what may be considered as a translation of Darwin's theory to thermodynamics. Biological systems have many possibilities for moving away from thermodynamic equilibrium and it is important to know along which pathways among the possible ones a system will develop. This leads to the following hypothesis sometimes denoted as the ecological law of thermodynamic (ELT) [8, 9, 16, 17, 24–26]: **If a system receives an input of exergy, then it will utilize this exergy to perform work. The work performed is first applied to maintain the system (far) away from thermodynamic equilibrium whereby exergy is lost by transformation into heat at the temperature of the environment. If more exergy is available, then the system is moved further away from thermodynamic equilibrium, reflected in growth of gradients. If more than one pathway is offered to depart from equilibrium, then the one yielding the highest eco-exergy storage (denoted Ex) will tend to be selected. Or expressed differently: among the many ways for ecosystems to move away from thermodynamic equilibrium, the one maximizing dEx/dt under the prevailing conditions will have a propensity to be selected.**

This hypothesis is supported by several ecological observations and case studies [16, 18]. Survival implies maintenance of the biomass and growth means increase of biomass and information. It costs exergy to construct biomass and gain information, and biomass and information therefore possess exergy. Survival and growth can, therefore, be measured by use of the thermodynamic concept eco-exergy, which may be understood as the work capacity the ecosystem possesses.

The idea of SDMs is to find continuously a new set of parameters (limited for practical reasons to the most crucial, i.e., sensitive parameters) that is better fitted for the prevailing conditions of the ecosystem. "Fitted" is defined in the Darwinian sense by the ability of the species to survive

Table 7.2: The β-values.

Early organisms	Plants		Animals
Detritus		1.00	
Virus		1.01	
Minimal cell		5.8	
Bacteria		8.5	
Archaea		13.8	
Protists Algae		20	
Yeast		17.8	
		33	Mesozoa, Placozoa
		39	Protozoa, amoeba
		43	Phasmida (stick insects)
Fungi, moulds		61	
		76	Nemertina
		91	Cnidaria (corals, sea anemones, jelly fish)
	Rhodophyta	92	
		97	Gastroticha
Prolifera, sponges		98	
		109	Brachiopoda
		120	Platyhelminthes (flatworms)
		133	Nematoda (round worms)
		133	Annelida (leeches)
		143	Gnathostomulida
	Mustard weed	143	
		165	Kinorhyncha
	Seedless vascula plants (incl. ferns)	158	
		163	Rotifera (wheel animals)
		164	Entoprocta
	Moss	174	
		167	Insecta (beetles, flies, bees, wasps, bugs, ants)
		191	Coleoidea (sea squirt)
		221	Lipidoptera (bufferflies)
		232	Crustaceans
		246	Chordata
	Rice	275	
	Gymnosperms (inl. pinus)	314	
		310	Mollusca, bivalvia, gastropodea
		322	Mosquito
	Flowering plants	393	
		499	Fish
		688	Amphibia
		833	Reptilia
		980	Aves (birds)
		2127	Mammalia
		2138	Monkeys
		2145	Anthropoid apes
		2173	*Homo sapiens*

and grow, which may be measured as discussed above by eco-exergy. Figure 7.3 shows the proposed modelling procedure, which has been applied in the development of totally 20 SDMs.

For all SDMs developed with eco-exergy as the goal function, the changes obtained were in accordance with actual observations. At least in models the applicability of the eco-exergy calculations have shown their more practical use, which may be can be explained by a robustness in the model calculations.

It is noteworthy that Coffaro *et al.* [27], in his SDM of the Lagoon of Venice, did not calibrate the model describing the spatial pattern of various macrophyte species such as *Ulva* and *Zostera*, but used exergy-index optimization to estimate parameters determining the spatial distribution of these species. He found good accordance between observations and model, as it was possible by this method *without* calibration, to explain more than 90% of the observed spatial distribution of various species of *Zostera* and *Ulva*.

Figure 7.4 illustrates the theoretical considerations behind the development of a SDM with eco-exergy as the goal function.

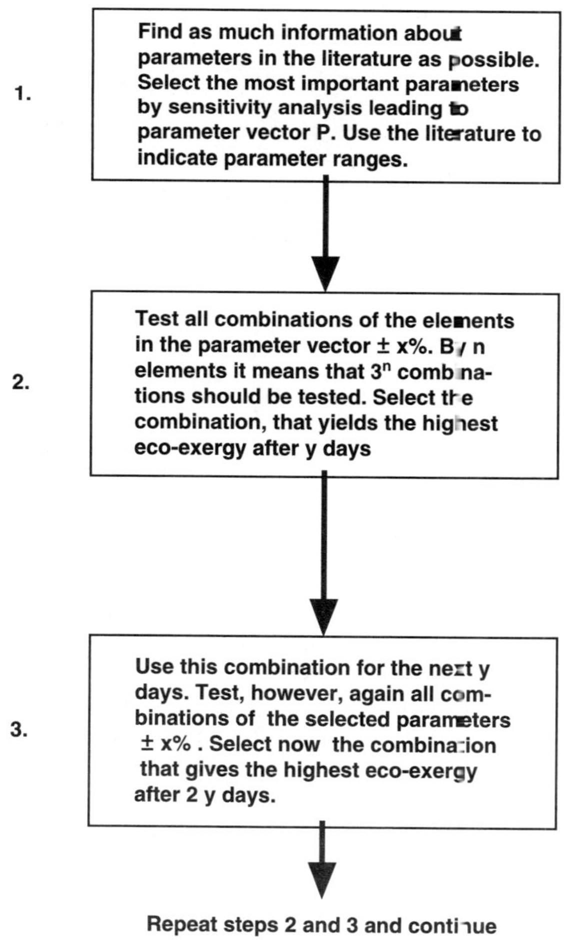

Figure 7.3: The procedure used for the development of SDMs. x% is selected in accordance with the knowledge about the variability of the parameters and y days are selected in accordance with the knowledge to the dynamic of the changes in the ecosystem.

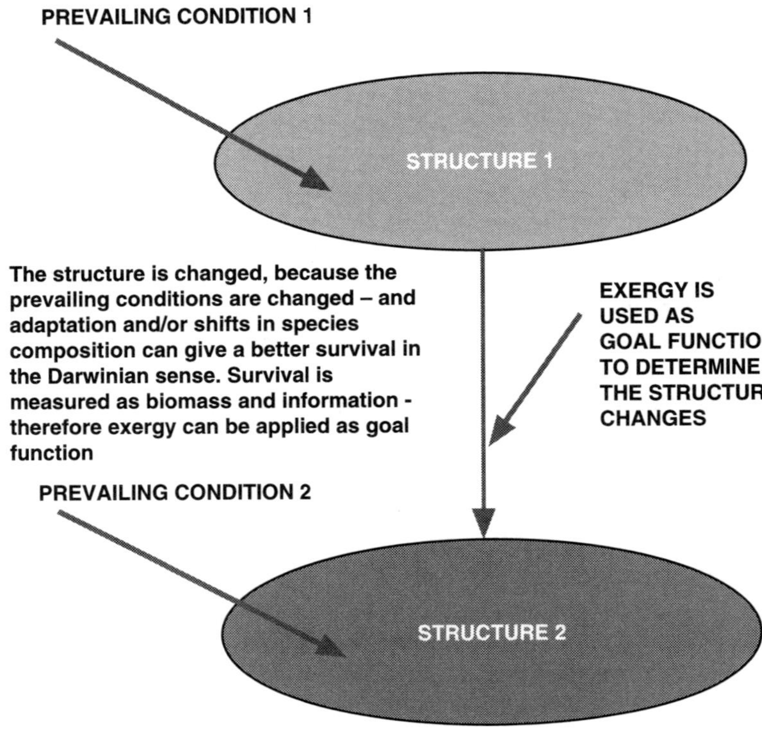

Figure 7.4: The theoretical considerations behind SDMs developed with eco-exergy as the goal function are illustrated.

Application of SDM is, of course, more cumbersome to apply than other models due to the eco-exergy optimization, which takes place for instance from every 5 to 30 days, but with a modern fast computer the additional computation is of course limited. The advantage of SDM is clear that eventually structurally dynamic changes are considered and if that is the case, a SDM will inevitably give a more accurate result. It may also be needed to use SDM for the calibration, because changes in, for instance, phytoplankton composition from spring to summer to fall may imply that different parameters should be applied in the different seasons. The conclusion is, therefore, that it is recommended to use SDM whenever it is known or even suspected that structural changes will take place. SDM requires, however, good observations, in most cases also of some structural changes to give acceptable results. A SDM will not necessarily be more expensive to develop than other models, but the need for good observations and a good data set will often make the entire project more expensive.

As mentioned above, eco-exergy has been applied to develop SDMs in 20 cases [7–9]. The 20 case studies are:

1–8. eight eutrophication models of six different lakes,

9. a model to explain the success and failure of biomanipulation based on removal of planktivorous fish,

10. a model to explain under which circumstances submerged vegetation and phytoplankton are dominant in shallow lakes,

11. a model of Lake Balaton which was used to support the intermediate disturbance hypothesis,
12–14. three small population dynamic models,
15. a eutrophication model of the Lagoon of Venice,
16. the Mondego Estuary and
17. an ecotoxicological model focusing on the influence of copper on zooplankton growth rates.
The last three case studies are presented below in Sections 7.5, 7.6 and 7.8 as illustrations of structurally dynamic modelling,
18. a model of Darwin's finches,
19. a model of the interaction between parasites and birds and
20. the SDM included in Pamolare 1 applied on Lake Fure in Denmark.

The tenth case is presented in Section 7.7 and is included to illustrate how STELLA can be used to develop a SDM and how the calibration can be improved by use of the structurally dynamic approach.

7.5 Development of SDM for Darwin's finches

The development of SDM for Darwin's finches (see details in Jørgensen and Fath [17]). The model reflects — as all models — the available knowledge which in this case is comprehensive and sufficient to validate even the ability of the model to describe the changes in the beak size as a result of climatic changes, causing changes in the amount, availability, and quality of the seeds that make up the main food item for the finches. The medium ground finches, *Geospiza fortis*, on the island Daphne Major were selected for this modelling case due to a very detailed case specific information found in Grant [28]. The model has three state variables: seed, Darwin's finches adult and Darwin's finches juvenile. The juvenile finches are promoted to adult finches 120 days after birth. The mortality of the adult finches is expressed as a normal mortality rate + an additional mortality rate due to food shortage and an additional mortality rate caused by a disagreement between bill depth and the size and hardness of seeds.

The beak depth can vary between 3.5 and 10.3 cm as pointed out by Grant [28]. The beak size $= \sqrt{DH}$,where D is the seed size and H is the seed hardness which are both dependent on the precipitation, particularly in the months January–April [28]. It is possible to determine a handling time for the finches for a given $\sqrt{DH}$ as function of the bill depth [28] which explains that the accordance between $\sqrt{DH}$ and the beak depth becomes an important survival factor. The relationship is used in the model to find a function called "diet" which is compared with $\sqrt{DH}$ to find how well the bill depth fits to the $\sqrt{DH}$ of the seed. This fitness function is based on the information given by Grant [28] about the handling time. It influences as mentioned above the mortality of adult finches, but also has an impact on the number of eggs laid and the mortality of the juvenile finches. The growth rate and mortality of seeds is dependent on the precipitation which is a forcing function known as function of time. A function called shortage of food is calculated from the food required for the finches which are known [28] and from the food available (the seed state variable). How the food shortage influences the mortality of juvenile finches and adult finches can be found in Grant [28]. The seed biomass and the number of *G. fortis* as function of time from 1975 to 1982 are known according to Grant [28]. These numbers from 1975 to 1976 have been used to calibrate the following parameters:

1. The influence of the fitness function on (i) the mortality of adult finches, (ii) the mortality of juvenile finches and (iii) the number of eggs laid.

2. The influence of food shortage on the mortality of adult and juvenile finches is known [28]. The influence is therefore calibrated within a narrow range of values.
3. The influence of precipitation on the seed biomass (growth and mortality).

All other parameters are known from the literature.

The exergy density is calculated (estimated) as 275 × the concentration of seed + 980 × the concentration of Darwin's finches (see Table 7.2). Every 15 days, it is found that a feasible change in the beak size, the generation time and the variations in the beak size if taken into consideration will give a higher exergy. If it is the case, then the beak size is changed accordingly. The modelled changes in the beak size were confirmed by the observations. The model results of the number of Darwin's finches are compared with the observations in Fig. 7.5. The standard deviation between modelled and observed values was 11.6% and the correlation coefficient (r^2) for modelled versus observed values is 0.977. The results of a non-SDM would not be able to predict the changes in the beak size and the finches would die off, but as the observations showed about 300 finches survived (see Fig. 7.5) due to the adaptation of the beak size, which is in accordance with the SDM too.

7.6 Model of the ectoparasites–birds interactions

Ectoparasites use in their interactions with birds a relatively high initial but thereafter decreasing mortality [29]. The results are also consistent with what is found by other parasite observations by Tripet and Richner [29]. Why do ectoparasites use this strategy? It can be shown by a SDM that the strategy is consistent with an optimization of eco-exergy for the parasites.

The model has three state variables: number of birds (hosts), number of parasites and weight of parasites.

The differential equations of the state variables are:

$$\text{dbird/dt} = -\text{mortality}$$

$$\text{dparasites/dt} = -\text{mortality of parasites}$$

$$\text{dweight/dt} = \text{growth rate}$$

The model follows over one season; let us say 120 days, the survival of a number of birds that have been infected by ectoparasites. Therefore, the number of birds and ectoparasites are decreasing steadily, while the ectoparasites are growing – the weight is increasing for the individual ectoparasites over time. The ectoparasites use a strategy whereby they obtain the highest survival which is also or at least close to the highest survival of the host, because the survival of the ectoparasites is dependent on the survival of the host. On the other side, the host will inevitably reduce the probability of survival due to the presence of the ectoparasites. The ectoparasites find a balance, which can be shown to correspond to a relatively high initial but thereafter decreasing mortality of the parasites. By use of a SDM with eco-exergy of the parasites as model goal function, it should be possible to find the optimum strategy of the growth rate and the mortality rate of the ectoparasites.

The eco-exergy is calculated as contribution of eco-exergy from the birds and contribution of eco-exergy from the parasites. The contribution of eco-exergy from the birds is

$$\text{Eco-exergy of birds} = \text{birds} \times 20 \times 980,$$

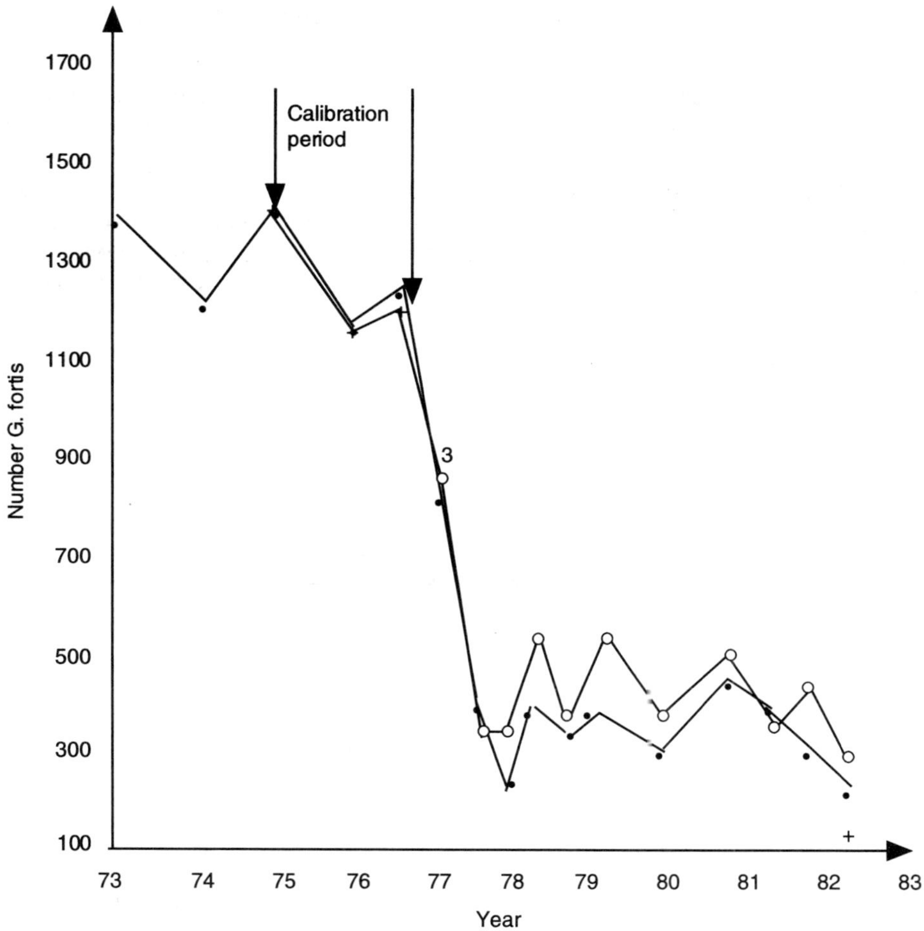

Figure 7.5: The observed number of finches (•) from 1973 to 1983 compared with the simulated result (○). For calibration, 75 and 76 were used and for the validation 77 and 78.

corresponding to the weight (20 g) and the information content factor 980 (see Table 7.2) and the contribution of eco-exergy from the parasites is

Eco-exergy of the parasites = number of parasites × weight of parasites × 167,

where 167 is the information content factor for the parasites (see again Table 7.2).

Total eco-exergy = eco-exergy of the birds + eco-exergy of the parasites.

Figure 7.6 gives that conceptual diagram using the symbols of the STELLA software.

From the differential equations it can be seen that there are three processes. They are formulated by use of the following equations:

Mortality (birds) = bird mortality coefficient × number of parasites × weight of parasites × number of birds

Mortality parasites = 0.025 × mortality coefficient × number of parasites/(number of birds + 1)$^{0.3}$

Growth rate (of parasites) = (basic growth rate + growth rate increase) × (1 – weight/max. weight)

Before it is possible to run the model it is necessary to select the initial values of the state variables, the parameters and the initial value of the two parameters that are presumed functions of the time.

The model has been developed in a C++ program, which is enclosed on a CD in the *Handbook of Ecological Modelling and Informatics* by Jørgensen *et al.* [30]. STELLA is not able to make a current optimization of a goal function. Use of STELLA for the development of SDMs is possible but cumbersome. Section 7.7 will, however, present an illustration where STELLA has been used to develop a structurally dynamic eutrophication model.

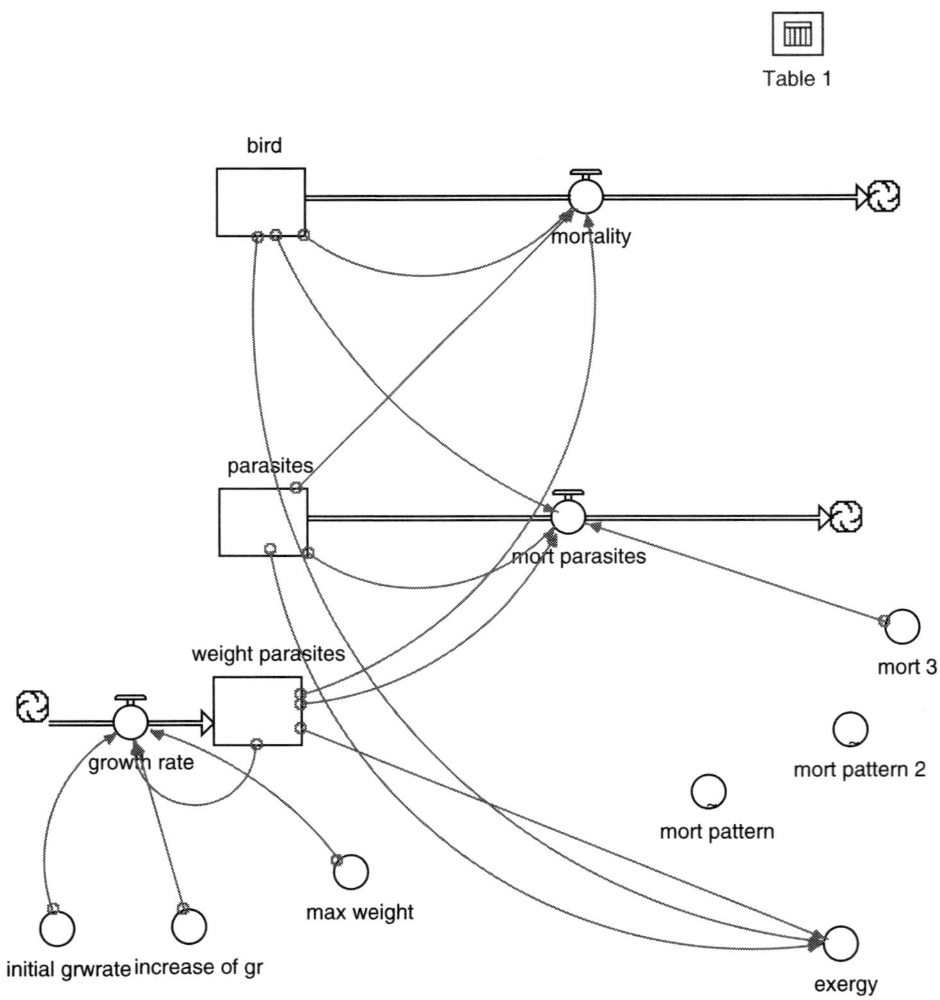

Figure 7.6: The conceptual diagram of the model using the symbols of the STELLA software. The model has been developed in C++ because STELLA is not able to optimize a goal function – in this case exergy. The C++ model is available on the enclosed CD in the *Handbook of Ecological Modelling and Informatics* by Jørgensen *et al.* [30].

For the parasite–bird SDM, it cannot be excluded that some parameter combinations will not give feasible results, if they are chosen randomly. Therefore, it is recommendable to use the default values (ranges and default values; see Table 7.3) when the parasite–bird model is developed and applied.

It is recommended to run the model simulation for a period of 60–180 days, for instance, 120 days equal to 4 months. When the SDM approach is applied, the model will stop every x days selected by the modeller. Two parameters namely mortality coefficient of the parasites and the growth rate increase of parasites are functions of the time.

When the model is stopped every x days, all combinations of the present mortality coefficient + and – 0.02 and the present growth rate increase + and – 0.001, totally nine combinations, are tested for the next x days and the combination giving the highest total eco-exergy after the x days is selected and so on. It is possible in C++ to carry this optimization procedure automatically, but not in STELLA.

The core model results are the three state variables and the eco-exergy of the birds, the eco-exergy of parasites, which is applied for the optimization, and the total eco-exergy. All variables are indicated as function of time by the software available on the CD in the *Handbook of Ecological Modelling and Informatics*, using both graphs and tables.

If we run the model, for instance, with an initial mortality of 0.5 1/24 h the SDM will show a stepwise decrease in the mortality and a stepwise increase in the growth rate. Let us say that the optimization was carried out every 10 days, it means totally 12 times. The mortality will then decrease from the initial value, 0.5 1/24 h, to 0.26 1/24 h, in average 0.38 1/24 h. Compare the final eco-exergy of the birds, of the parasites and the total eco-exergy with the non-structurally dynamic case, where an average mortality (in our example 0.38 1/24 h) and average growth rate are used. The comparison will show that the structurally dynamic case, which gives the possibility for variable parameters, is resulting in higher eco-exergy values, both for the parasites, the birds and the total eco-exergy. The strategy to start at a higher but decreasing mortality and a lower but increasing growth rate is beneficial for the parasites and the birds.

It is, of course, possible to apply different mortality and growth rates and they will all give higher total eco-exergy for the structurally dynamic approach compared with the constant parameter case. The structurally dynamic approach opens for the possibilities to vary the parameters if the result is a higher eco-exergy.

Figure 7.7 illustrates the results of several model run, using different initial values for the mortality rate. The eco-exergy for a constant mortality rate is much lower (named Eco-exergy

Table 7.3: Parameters and initial values of the state variables.

	Possible range	Default value
Initial number of birds	50–10000	100
Initial number of parasites	100–1000	240
Initial values of weight (g)	0.001–0.2	0.02
Bird mortality coefficient	0.000001–0.001	0.00001
Initial value of mortality coefficient	0.005–1.0	0.5
Basic growth rate	0.0001–0.1	0.005
Growth rate increase, initial value	–0.01 – +0.05	0.001
Max. weight	0.05–0.3	0.12

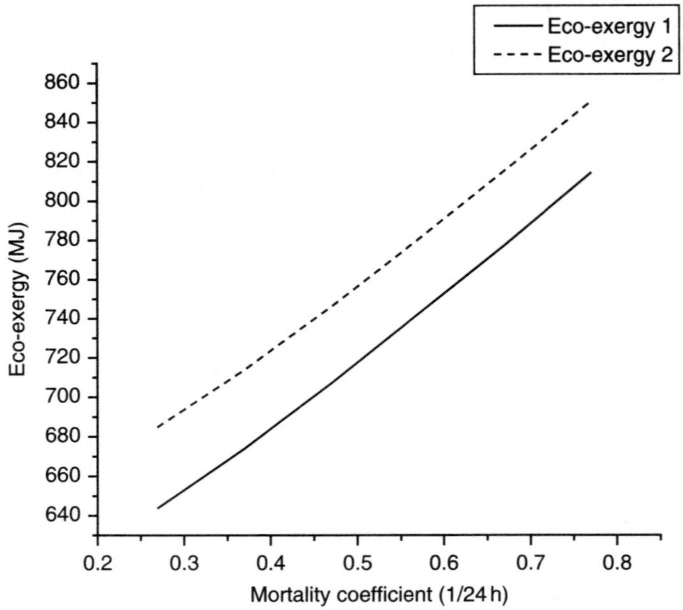

Figure 7.7: Eco-exergy for different mortality coefficients is shown. The lower line corresponds to a constant mortality coefficient while the upper line represents the eco-exergy for the average value, but starting at a mortality coefficient that is 0.23 higher than the average value and finishing with a mortality coefficient that is 0.23 lower than the average value.

1 in the figure) than the eco-exergy using a decreasing mortality rate (Eco-exergy 2) with an average value equal to the constant values for the Eco-exergy 1 run.

7.7 A structurally dynamic eutrophication model developed by use of STELLA

Zhang *et al.* [31, 32] have developed SDMs by use of STELLA, although as mentioned above it is cumbersome procedure to introduce the application of a goal function by development of a STELLA model. The conceptual diagram of the model is shown in Fig. 7.8. The model was developed by use of the data from Lake Mogan close to Ankara, Turkey. Phosphorus is the limiting factor for eutrophication in the lake, which is interesting because it is a shallow lake with a competition between phytoplankton and submerged vegetation. The model has seven state variables: soluble P (PS), phosphorus in phytoplankton (PA), phosphorus in zooplankton (PZ), phosphorus in detritus (PD), phosphorus in submerged plants (PSM), exchangeable phosphorus in the sediment (PEX) and phosphorus in pore water (PP). The processes are inflows and outflows of phosphorus, phosphorus in phytoplankton and phosphorus in detritus. Soluble phosphorus is taken up by phytoplankton – the process is named uptakeP. Zooplankton grazes on phytoplankton indicated as grz on the diagram. The settling of detritus and phytoplankton is covered by the same method as indicated for the eutrophication model presented in Chapter 3. A part of the settled material is lost as non-exchangeable phosphorus, while the exchangeable fraction is going to the state variable

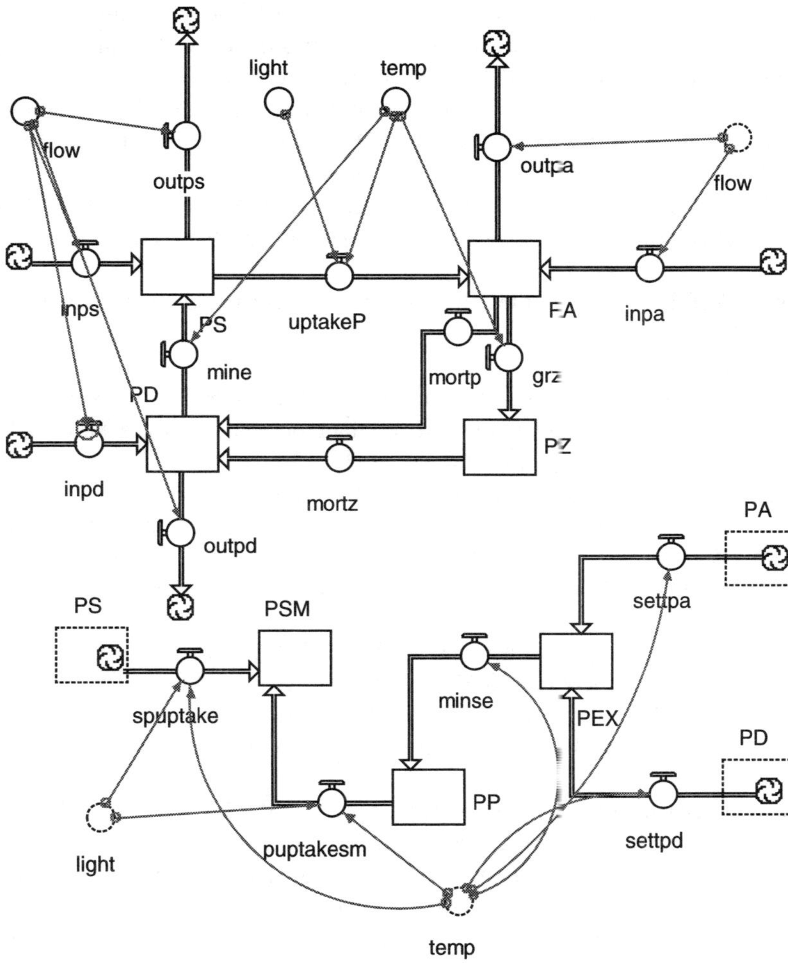

Figure 7.8: The conceptual diagram of the Lake Mogan eutrophication model focusing on the cycling of phosphorus. The model has seven state variables: soluble P (PS), phosphorus in phytoplankton (PA), phosphorus in zooplankton (PZ), phosphorus in detritus (PD), phosphorus in submerged plants (PSM), exchangeable phosphorus in the sediment (PEX) and phosphorus in pore water (PP).

exchangeable phosphorus (PEX). A mineralization of the exchangeable phosphorus takes place in the sediment – the process is named minse in the diagram.

Mineralization in the water phase of detritus phosphorus is a process named mine. Temperature influences all the process rates. Light is, of course, considered as a climatic forcing function influencing the growth of both phytoplankton and submerged plants. The submerged plants take most phosphorus up from the sediment, but the model considers both uptake of phosphorus from water and sediment by the submerged plants.

The SDM approach was used in the presented eutrophication model to examine:

1. The possibilities to improve calibration. Usually eutrophication models are using one set of parameters for the entire annual cycle, but almost all lakes have different species of

phytoplankton and zooplankton in the spring, in the summer and in the fall, because the conditions are different from season to season. The question is if we could include these changes of the parameters corresponding to the shifts in species composition from season to season, would we then obtain a better calibration and validation? In other words, could we improve calibration and validation if we would use the SDM-approach not only for the development of prognoses but also for the calibration and validation?
2. Whether the catastrophic changes from submerged vegetation dominance to phytoplankton dominance that is described by Scheffer *et al.* [33] could be covered by the SDM.

The use of the SDM approach in the calibration phase was carried out by use of the following stepwise method:

I. To reduce the number of parameter combination, the following allometric relationships between parameters of phytoplankton and zooplankton and their sizes were applied:
   a. Max. growth rate of phytoplankton = $3.0 - 0.3 \times$ LVP,
   b. Max. growth rate of zooplankton = $0.72 - 0.13 \times$ LVZ,
   c. Michaelis Menten half-saturation constant for grazing by zooplankton: KA = $-0.15 + 0.66 \times$ LVZ,
   d. Michaelis Menten half-saturation constant for phosphorus uptake by phytoplankton: KP $= 0.0005 \times$ LVP,
   e. Mortality rate of phytoplankton = $(3.0 - 0.3 \times \text{LVP}) \times 0.1/3$,
   f. Mortality rate of zooplankton = $0.050 - 0.001 \times$ LVZ,
   g. Settling rate of phytoplankton = $0.05 + 0.006 \times$ LVP,

   where, LVP and LVZ are the logarithm of sizes for phytoplankton and zooplankton (μm^3), respectively.

   In accordance to the procedure for the development of SDMs (see Fig. 7.3) the model should be tested for all combinations of at least three possible values of the variable parameters. It means that if seven parameters are made structurally dynamic as it was decided for the eutrophication model in this case, it is required to run the model 3^7 times = 2187, but if as it is done by use of the allometric principles it is possible to reduce the parameters to two – namely size of phytoplankton and size of zooplankton – it is only required to run the model 3×3 times or 9 times.
II. First, the model was calibrated using the usual trial-and-error method to find the combinations of parameters that would give the smallest discrepancy between model results and observations. These parameters were maintained for the subsequent application of the structurally dynamic approach, except for the seven parameters listed above – they would be determined by a current change of the size according to the SDM-approach, described as point 3.
III. Nine runs (3 phytoplankton sizes – the same , 10% increase and 10% decrease and 3 zooplankton sizes – the same , 10% increase and 10% decrease) are performed from day 0 to day x (x = 10 days was chosen for the Lake Mogan model). The combination of size for phytoplankton and zooplankton that gave the highest eco-exergy value was selected for the 10 days. Eco-exergy was calculated for the model as Eco-exergy = $21 \times$ phytoplankton + $135 \times$ zooplankton + $100 \times$ submerged vegetation + detritus. This procedure was repeated every 10 days. After an annual model run, it was possible to give a graph or table of the phytoplankton and zooplankton sizes, which would currently – it means every 10 days – optimize the eco-exergy of the model. The sizes were "translated" to the seven parameters that were selected as structurally dynamic – see the list above (a–g).

IV. The size of phytoplankton and zooplankton, which was found as function of time, was introduced to the model and the other model parameters, that is, all the parameter minus the seven phytoplankton and zooplankton parameters, were now calibrated again to account in this calibration for the influence of our knowledge about the phytoplankton and zooplankton size as function of time. In principle, III and IV should be repeated until there are no further changes of the parameters obtained.

This stepwise procedure (see the steps 1–3 in Fig. 7.9) is of course more cumbersome than to use an automatic SDM programmed in C++, but still it takes only a few days to obtain this SDM-calibration, while the conventional calibration by trial and error may even require much longer time. The procedure II–IV is represented as steps 1–3 schematically in Fig. 7.9. For the Lake Mogan model, it was found that the standard deviation expressing the difference between the modelled value and observed value of phytoplankton phosphorus for the SDM-calibration was 10.9% versus 18% for the usual calibration procedure. For phytoplankton, zooplankton and soluble phosphorus, which were considered the most important state variables, the standard deviation of 8.2% for the SDM-calibration and 10.7% for the usual calibration. The graphs for the observations and the model outputs of the two different calibration procedures are shown in Fig. 7.10.

The SDM was used after calibration to answer the following question: is it possible to describe the catastrophic changes from submerged vegetation dominance to phytoplankton dominance

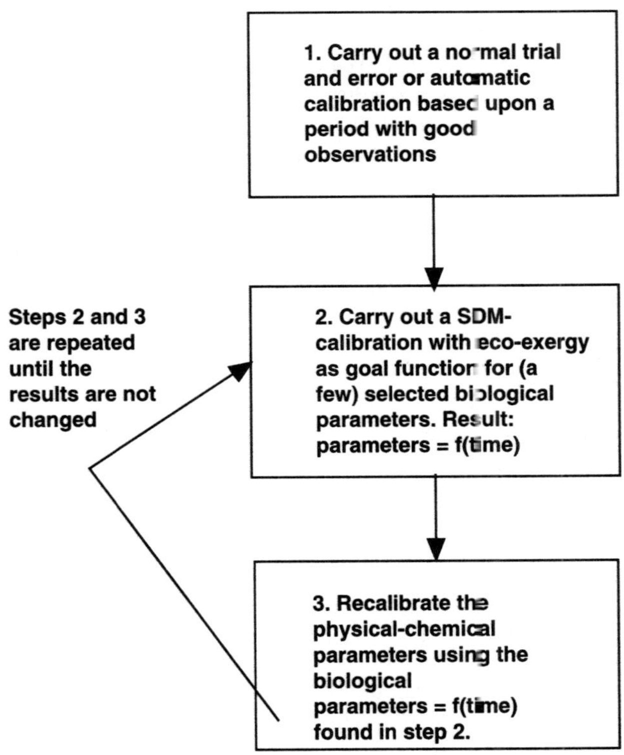

Figure 7.9: The diagram shows how the structurally dynamic approach is applied for an improved calibration procedure. The steps 2 and 3 are applied iteratively until the results of the calibration are not changed further.

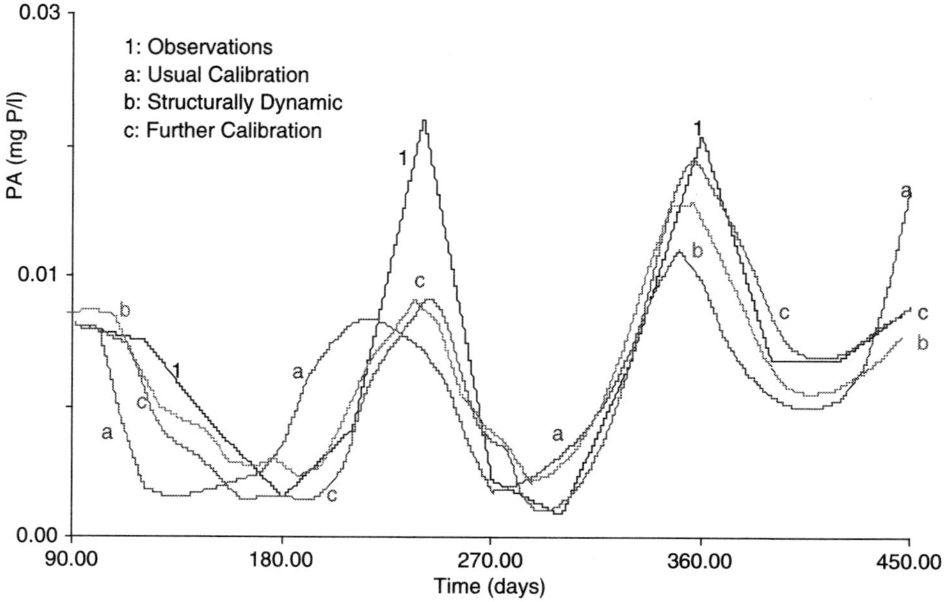

Figure 7.10: The phytoplankton-P as mg/l is shown for (1) observations: (a) the calibration obtained by the conventional trial-and-error calibration; (b) the calibration of the SDM (for the method see the text); (c) the calibration obtained after the non-structurally dynamic parameters have been calibrated by use of the structurally dynamic parameters obtained by procedure (b). The time is from the first of October (day 90) to first of October (day 450). The results of the final calibration, SDM calibration followed by the normal calibration of the non-structurally dynamic parameters (c) are very well in accordance with the observation, except for the first peak in late April. The deviation between the modelled and the observed phytoplankton concentration at the second peak in July is only 8% relatively, while the trial-and-error method here gives a deviation of almost 40%.

and back again, which is described by Scheffer *et al.* [33], by application of the SDM of Lake Mogan? In accordance with Scheffer *et al.* we should expect that submerged vegetation is replaced by phytoplankton if we increase the phosphorus concentration to 250 μg/l and that the submerged vegetation does not return when we decrease the phosphorus concentration before at about 100 μg/l. Can we in other words simulate the described hysteresis behavior by use of the developed SDM for Lake Mogan? The phosphorus concentration in the lake is about 80–85 μg/l according to the observations and it was the concentration applied for the calibration described above.

To answer the question, the phosphorus concentration in the water was increased a factor 5 times which implies that we should expect a phosphorus concentration after a couple of years corresponding to the retention time of about 400 μg/l. Afterwards the phosphorus concentration was reduced to the present level of 80–85 μg/l. Figure 7.11 illustrates the results of these changes in the phosphorus concentration from 80–85 μg/l to 400 μg/l and from 400 μg/l back to 80–85 μg/l.

Figure 7.12 shows the phytoplankton-P concentrations as function of time for five different phosphorus concentrations of the water flowing into the lake: 1) 0.5 times the present level, 2)

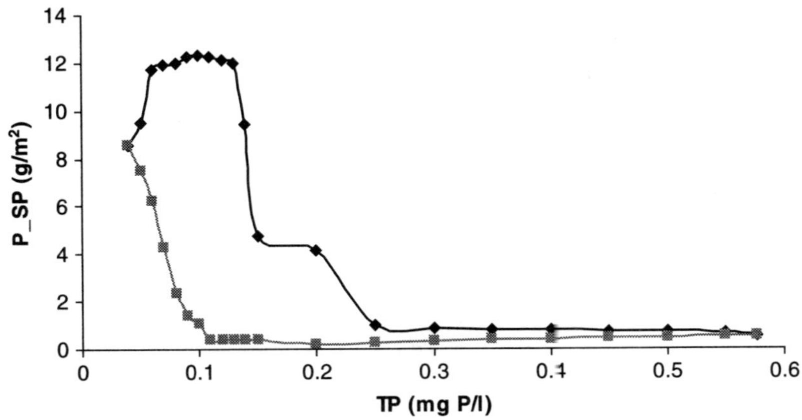

Figure 7.11: The graphs show the reaction of submerged plants phosphorus to an increase of phosphorus to about 400 μg/l followed by return to the original about 80–85 μg/l. When the phosphorus concentration is increased P-SP is increasing, but at about 250 μg/l the submerged vegetation disappear and is replaced by phytoplankton-P. At the return to the 80–85 μg/l, the submerged vegetation emerged at 100 μg/l. The hysteresis behavior is completely in accordance with Scheffer *et al.* [33].

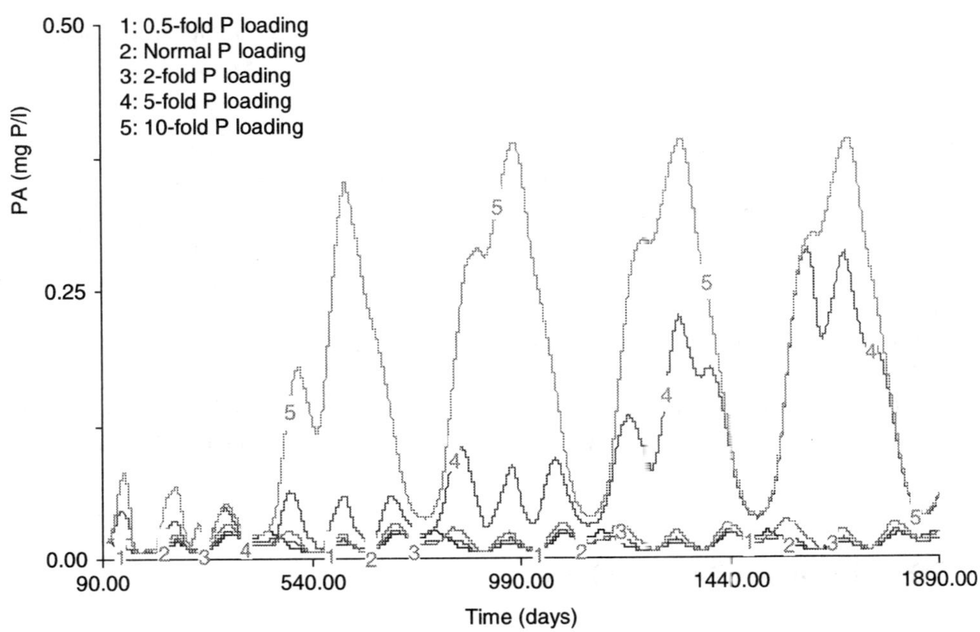

Figure 7.12: The phytoplankton-P concentration as function of time is shown when the phosphorus concentration of the water flowing into the lake is increased by 1) a factor 0.5, 2) normal factor 1.0, 3) a factor 2, 4) a factor 5 and 5) a factor 10. The present concentration of phosphorus in the lake is 80–85 μg/l. 1), 2) and 3) give no changes, while 4) and 5) give a significant increase of phytoplankton, which becomes dominant.

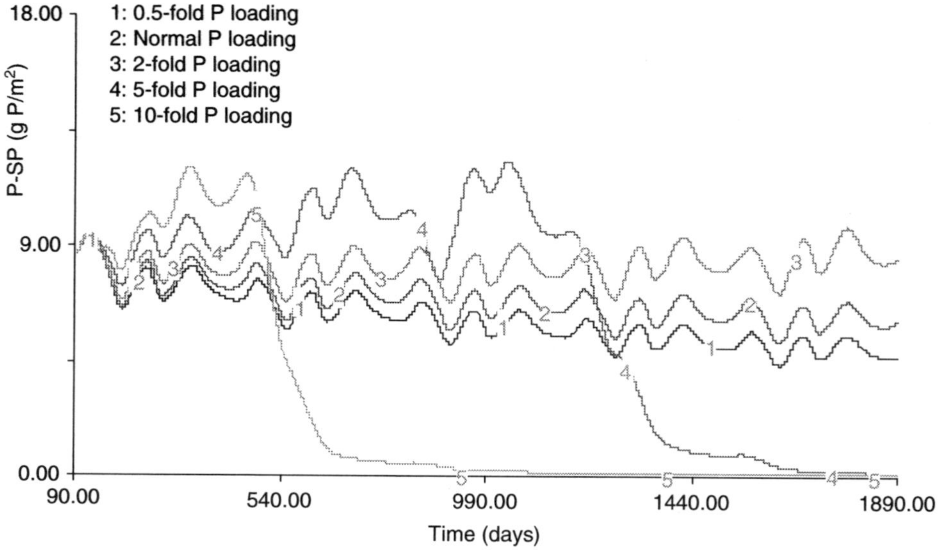

Figure 7.13: The submerged plant concentration as gP/m^2 as function of time is shown when the phosphorus concentration of the water flowing into the lake is increased by 1) a factor 0.5, 2) normal factor 1.0, 3) a factor 2, 4) a factor 5 and 5) a factor 10. The present concentration of phosphorus in the lake is 80–85 μg/l. 1), 2) and 3) give no changes, while 4) and 5) first give a minor increase of the concentration followed by a significant decrease when phytoplankton becomes dominant, compare with Fig. 7.12

the present level, 3) two times the present level, 4) five times the present level and 5) ten times the present level. As indicated above, the present level is 80–85 μg P/l.

Figure 7.13 shows the submerged plant as g/m^2 as function of time for the same five phosphorus concentrations of the water flowing into the lake. The shift in dominance from submerged vegetation to phytoplankton is very clear for the phosphorus concentrations five and ten times the present value.

It is possible to conclude from the presented application of SDM that

1. use of the SDM-approach for the calibration is able to improve the calibration results and
2. that it is possible to answer the raised questions about the catastrophic changes by use of SDM.

7.8 SDM developed for Lake Fure

Lake Fure is the deepest lake in Denmark. It has a surface area of 941 ha and an average depth of 13.5 m. It consists of two ecologically different parts that are connected: the main basin, which is the large and deep part with an average depth of 16.5 m and the maximum depth of 37.7 m, and the shallow small part, which is called Store Kalv and has a mean and a maximum depth of 2.5 m and 4.5 m, respectively. The lake is situated 12–17 km from the center of Copenhagen and has therefore great recreational value.

The eutrophication of the lake increased significantly in the 1950s and 1960s due to growing populations in the suburbs north of Copenhagen. The wastewater was treated mechanically and biologically but there was no removal of nutrients. It was therefore decided in the late 1960s to take measures to reduce the eutrophication. In 1970, it was decided that the three municipalities discharging wastewater into the lake could either choose to treat the wastewater effectively by removal of nutrients or pump the wastewater into the sea. A maximum phosphorus concentration of 0.2 mg/l and a maximum nitrogen concentration of 8 mg/l were required for the treated wastewater if the municipality would select the treatment solution. One of the three municipality selected the treatment solution, while the two other municipality preferred to pump the mechanically and biologically treated water into the sea. A model developed at that time showed that the treatment solution was the best solution for the lake due to a faster recovery of the lake. The lake had water retention of approximately 16 years and the pumping of round 2 million m^3 into the sea prolonged the retention to about 21 years, which implied, of course, a slower reduction of the eutrophication.

The nutrient balances of the lake in 1972 just before the implementation of the above mentioned solution are shown in Fig. 7.14. The measures taken reduce the phosphorus and nitrogen discharge into the lake by wastewater to less than 1 t and about 3 t respectively. The three other sources of nutrients, drainage water, storm water overflow and precipitation were unchanged. During the period 1972–2000, the municipalities were able to reduce the other sources slightly as it can be seen in Fig. 7.15 showing the nutrient balances for the year 2000. The nutrient input from storm water was reduced by enlarging the storm water capacity. Notice, however, that the internal load-

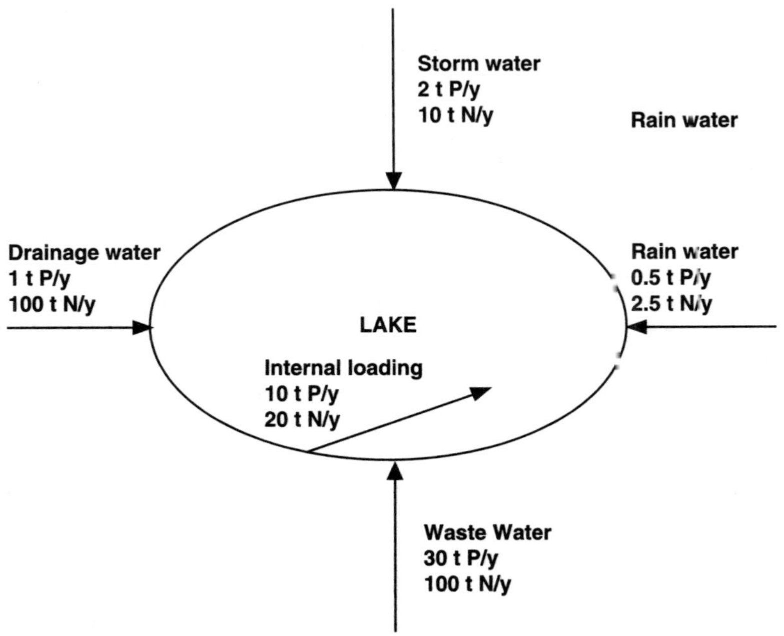

Figure 7.14: The nutrient balances in the year 1972 before the loading of nutrients from wastewater was reduced.

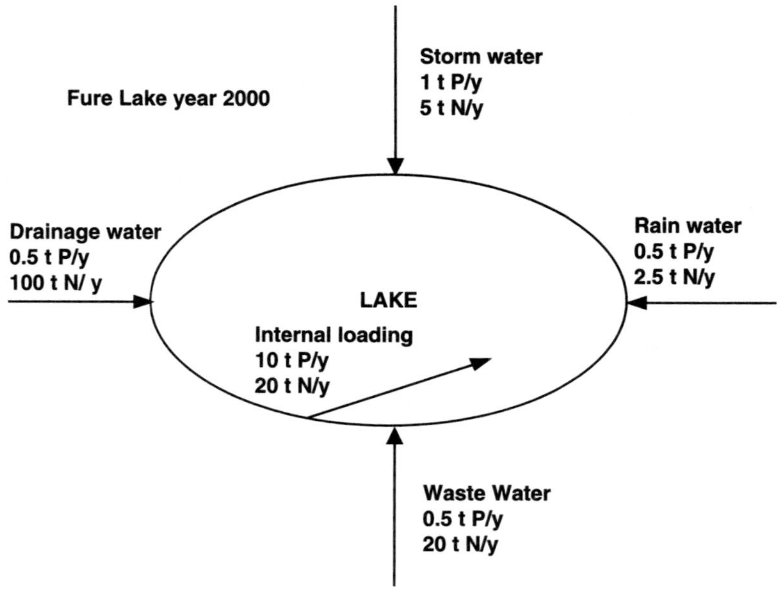

Figure 7.15: The nutrient balances in the year 2000.

ing was not reduced which is at least partly due to long retention time. It belongs to the story that the two municipalities that preferred to pump the wastewater into the sea in the early 1970s and did not accept the model result that the treatment solution would give a faster recovery, were forced in 1986 to include nitrogen and phosphorus removal in the wastewater treatment due to introduction of maximum standard for discharge of all wastewater, even wastewater discharge directly into the sea.

Shortly after the year 2000 it was decided to use restoration methods to recover the lake faster. Two restoration methods were proposed:

1. Aeration of the hypolimnion by oxygen from late April to late October when the lake had a thermocline. The release of phosphorus from the sediment would thereby be reduced significantly.
2. Biomanipulation by massive removal of planktivorous fish.

The restoration of the lake started in 2003. A SDM was developed in the year 2005 (see the paper by Gurkan *et al.* [34]). A SDM was natural to apply in this case as structural changes were expected due to reduced internal loading, the removal of a significant part of the planktivorous fish and further increase of the storm water capacity. The model was used for prognosis about the water quality in the year 2010. Figure 7.16 shows the application of the model results on the nutrient balance and Table 7.4 shows the changes in the important state variables from 2004 to 2010 according to the prognosis made by the SDM model of Pamolare.

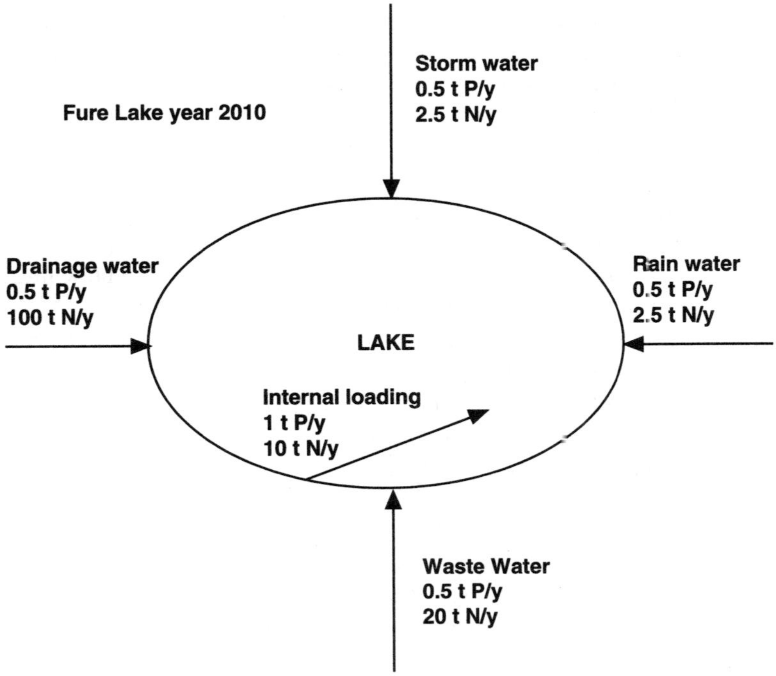

Figure 7.16: The nutrient balances according to the result of the SDM. The internal loading is reduced mainly due to the aeration of the hypolimnion and the phosphorus and nitrogen loading from storm water is reduced due to increase of the storm water capacity.

Table 7.4: Changes in the important state variables from the year 2000 to 2010 according to the results of the SDM applied on Lake Fure to assess the effect of the restoration project (aeration of hypolimnion, biomanipulation and increased storm water capacity).

State variable	Year 2000	Year 2010
Total P mg/l	0.18–0 25	0.02–0.07
Chl. a maximum (early Aug.) mg/l	0.044	0.022
Transparency minimum (early Aug.) m	2.2	3.6
Zooplankton mg d.w./l	0.8	1.8

References

[1] Wolfram, S., Cellular automata as models of complexity. *Nature*, **311**, pp. 419–424, 1984.

[2] Wolfram, S., Computer software in science and mathematics. *Sci. Am.*, **251**, pp. 140–151, 1984.

[3] Jørgensen, S.E., A holistic approach to ecological modelling by application of thermodynamics. *Systems and Energy*, eds W. Mitsch *et al.* Ann Arbor Press: Ann Arbor, MI, 1982.

[4] Jørgensen, S.E., Structural dynamic model. *Ecological Modelling*, **31**, pp. 1–9, 1986.

[5] Jørgensen, S.E., Use of models as experimental tools to show that structural changes are accompanied by increased exergy. *Ecological Modelling*, **41**, pp. 117–126, 1988.

[6] Jørgensen, S.E., Ecosystem theory, ecological buffer capacity, uncertainty and complexity. *Ecological Modelling*, **52**, pp. 125–133, 1990.

[7] Jørgensen, S.E., Parameters, ecological constraints and exergy. *Ecological Modelling*, **62**, pp. 163–170, 1992.

[8] Jørgensen, S.E., Development of models able to account for changes in species composition. *Ecological Modelling*, **62**, pp. 195–208, 1992.

[9] Jørgensen, S.E., *Integration of Ecosystem Theories: A Pattern*, 3rd edn, Kluwer Academic Publishing Company: Dordrecht, The Netherlands, 2002.

[10] Straskraba, M., Natural control mechanisms in models of aquatic ecosystems. *Ecological Modelling*, **6**, pp. 305–322, 1979.

[11] Straskraba, M., The effects of physical variables on freshwater production: analyses based on models. *The Functioning of Freshwater Ecosystems (International Biological Programme 22)*, eds E.D. Le Cren & R.H. McConnell, Cambridge University Press: Cambridge, pp. 13–31, 1980.

[12] Allen, T.F.H. & Starr, T.B., *Hierarchy, Perspectives for Ecological Complexity*. The University of Chicago Press: Chicago and London, 1982.

[13] Margalef, R.A., Networks in ecology. *Theoretical Studies of Ecosystems: The Network Perspectives*, eds M. Higashi & T.P. Burns, Cambridge University Press: Cambridge, pp. 41–57, 1991.

[14] Margalef, R.A., Information theory and complex ecology. *Compex Ecology*, eds B.C. Patten & S.E. Jørgensen, Prentice Hall PTR: New Jersey, pp. 40–50, 1995.

[15] Stenseth, N.C., Darwinian evolution in ecosystems: a survey of some ideas and difficulties together with some possible solutions. *Complexity. Language and Life: Mathematical Approaches*, eds J.L. Casti & A. Karlquist, Springer Verlag: Berlin, pp. 107–123, 1986.

[16] Jørgensen, S.E. & Fath, B.D., Application of thermodynamic principles in ecology. *Ecological Complexity*, **1**, pp. 267–280, 2004.

[17] Jørgensen, S.E. & Fath, B., *A New Ecology. A Systems Approach*, Elsevier: Amsterdam, 2007.

[18] Jørgensen, S.E. & Svirezhev, Y.M., *Towards a Thermodynamic Theory for Ecological Systems*, Elsevier: Amsterdam, 2005.

[19] Nielsen, S.N., *Application of Maximum Exergy in Structural Dynamic Models*, PhD Thesis, National Environmental Research Institute, Denmark, 1992.

[20] Nielsen, S.N., Strategies for structural-dynamical modelling. *Ecological Modelling*, **63**, pp. 91–102, 1992.

[21] Fontaine, T.D., A self-designing model for testing hypotheses of ecosystem development. *Progress in Ecological Engineering and Management by Mathematical Modelling. Proc. 2nd Int. Conf. State-of-the-Art Ecol. Mod.*, April 1980, Liège, Belgium, ed. D. Dubois, pp. 281–291, 1981.

[22] Patten, B.C., Synthesis of chaos and sustainability in a nonstationary linear dynamic model of the American black bear (Ursus americanus Pallas) in the Adirondack Mountains of New York. *Ecological Modelling*, **100**, pp. 11–42, 1997.

[23] Bossel, H., Real structure process description as the basis of understanding ecosystem. *Ecological Modelling*, **63**, pp. 261–276, 1992.

[24] Jørgensen, S.E. & Meyer, H.F., Ecological buffer capacity. *Ecological Modelling*, **3**, pp. 39–61, 1977.

[25] Jørgensen, S.E. & Mejer, H.F., A holistic approach to ecological modelling. *Ecological Modelling*, **7**, pp. 169–189, 1979.

[26] Fath, B., Jørgensen, S.E. & Bernard, C., Patten and Milan Straskaba. Ecosystem growth and development. *BioSystem*, **77**, pp. 213–228, 2004.

[27] Coffaro, G., Bocci, M. & Bendoricchio, G., Structural dynamic application to space variability of primary producers in shallow marine water. *Ecological Modelling*, **102**, pp. 97–114, 1997.

[28] Grant, P.R., *Ecology and Evolution of Darwin's Finches*, Reprinted in 1999, Princeton University Press: New Jersey, 1986.

[29] Tripet, F. & Richner, H., Density-dependent processes in the population dynamics of a bird ectoparasite, *Ceratophyllas gallinae*. *Ecology*, **80**, pp. 1257–1277, 2000.

[30] Jørgensen, S.E., Chon, T. & Recknagel, F., *Handbook of Ecological Modelling and Informatics*, WIT Press: Southampton, 2009.

[31] Zhang, J., Jørgensen, S.E., Tan, C.O. & Beklioglu, M., A structurally dynamic modelling – Lake Mogan, Turkey, as a case study. *Ecological Modelling*, **164**, pp. 103–120, 2003.

[32] Zhang, J., Jørgensen, S.E., Tan, C.O. & Beklioglu, M., Hysteresis in vegetation shift – Lake Mogan prognoses. *Ecological Modelling*, **164**, pp. 227–238, 2003.

[33] Scheffer, M., Carpenter, S., Foley, J.A., Folke, C. & Walker, B., Catastrophic change of ecosystems. *Nature*, **413**, pp. 591–596, 2001.

[34] Gurkan, Z., Zhang, J. & Jørgensen, S.E., Development of a structurally dynamic model for forecasting the effects of restoration of lakes. *Ecological Modelling*, **197**, pp. 89–103, 2006.

Exercises/problems – Chapter 7

1. Repeat from Chapter 2 the advantages and disadvantages of applying SDMs.
2. Why is not beneficial to apply the SDM-approach on the subsurface wetland model presented in Chapter 3?
3. Calculate the eco-exergy in kJ/m^2 of the shown steady state model, developed by the use of Ecopath. The boxes show the state variables in g/m^2 and the arrow shows the processes expressed in g/m^2 y.

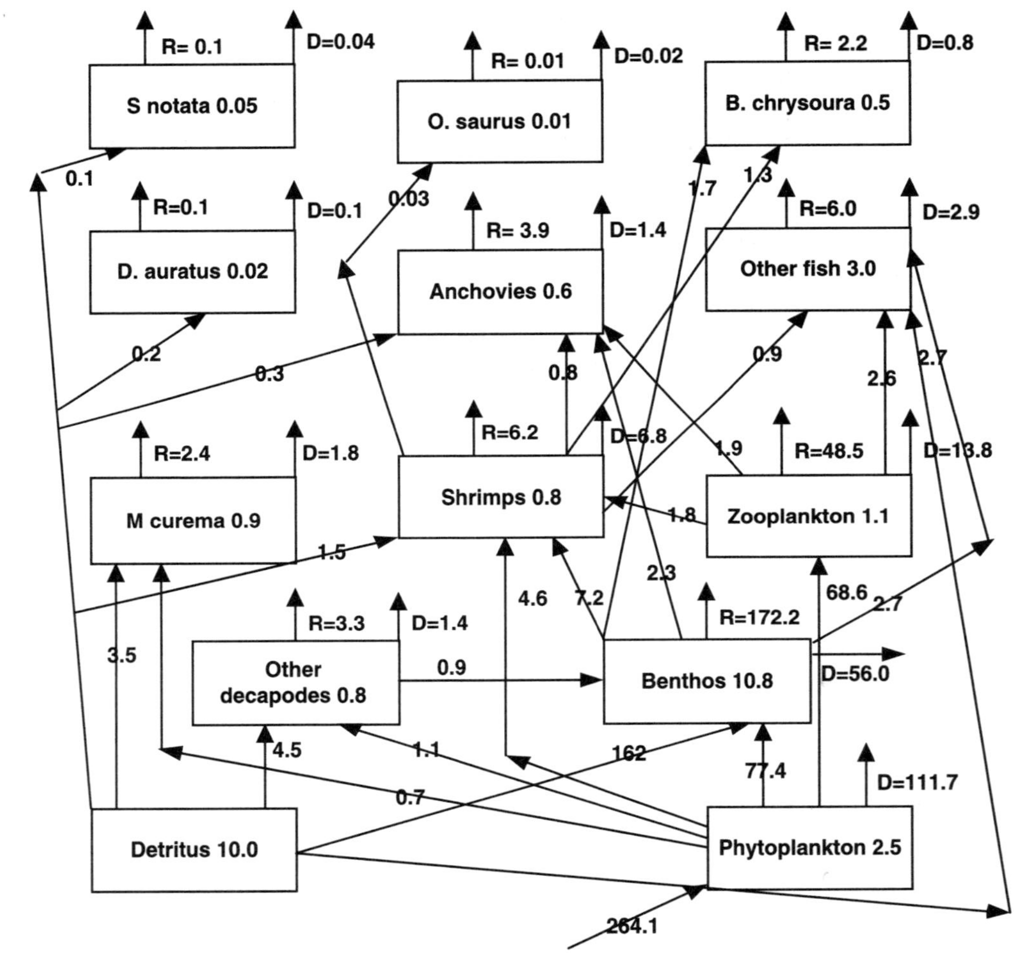

4. Explain why it is beneficial to apply the SDM-approach when we want to apply models to determine the ecosystem changes as a consequence of global warming.
5. Explain why it is considered a strong support for the ELT that the eco-exergy is approximately the same for the two possible structures, dominance by phytoplankton and dominance by submerged vegetation, in shallow lakes between about 100 μg and 250 μg.

Appendix 1: Table of oxygen solubility as function of temperature and salinity

Table A1: Dissolved oxygen (ppm, mg/l) in fresh, brackish and seawater at different temperatures and at different chlorinities (%). Values are at saturation.

Temp.	0%	0.2%	0.4%	0.6%	0.8%	1.0%	1.2%	1.4%	1.6%	1.8%	2.0%
1	14.24	13.87	13.54	13.22	12.91	12.58	12.29	11.99	11.70	11.42	11.15
2	13.74	13.50	13.18	12.88	12.56	12.26	11.98	11.69	11.40	11.13	10.86
3	13.45	13.14	12.84	12.55	12.25	11.96	11.68	11.39	11.12	10.85	10.59
4	13.09	12.79	12.51	12.22	11.93	11.65	11.38	11.10	10.83	10.59	10.34
5	12.75	12.45	12.17	11.91	11.63	11.36	11.09	10.83	10.57	10.33	10.10
6	12.44	12.15	11.86	11.60	11.33	11.07	10.82	10.56	10.32	10.09	9.86
7	12.13	11.85	11.58	11.32	11.06	10.82	10.56	10.32	10.07	9.84	9.63
8	11.85	11.56	11.29	11.05	10.80	10.56	10.32	10.07	9.84	9.61	9.40
9	11.56	11.29	11.02	10.77	10.54	10.30	10.08	9.84	9.61	9.40	9.20
10	11.29	11.03	10.77	10.53	10.30	10.07	9.84	9.61	9.40	9.20	9.00
11	11.05	10.77	10.53	10.29	10.06	9.84	9.63	9.41	9.20	9.00	8.80
12	10.80	10.53	10.29	10.06	9.84	9.63	9.41	9.21	9.00	8.80	8.61
13	10.56	10.30	10.07	9.84	9.63	9.41	9.21	9.01	8.81	8.61	8.42
14	10.33	10.07	9.86	9.63	9.41	9.21	9.01	8.81	8.62	8.44	8.25
15	10.10	9.86	9.64	9.43	9.23	9.03	8.83	8.64	8.44	8.27	8.09
16	9.89	9.66	9.44	9.24	9.03	8.84	8.64	8.47	8.28	8.11	7.94
17	9.67	9.46	9.26	9.05	8.85	8.65	8.47	8.30	8.11	7.94	7.78
18	9.47	9.27	9.07	8.87	8.67	8.48	8.31	8.14	7.97	7.79	7.64
19	9.28	9.08	8.88	8.68	8.50	8.31	8.15	7.98	7.80	7.65	7.49
20	9.11	8.90	8.70	8.51	8.32	8.15	7.99	7.84	7.66	7.51	7.36
21	8.93	8.72	8.54	8.35	8.17	7.99	7.84	7.69	7.52	7.38	7.23
22	8.75	8.55	8.38	8.19	8.02	7.85	7.69	7.54	7.39	7.25	7.11
23	8.60	8.40	8.22	8.04	7.87	7.71	7.55	7.41	7.26	7.12	6.99
24	8.44	8.25	8.07	7.89	7.72	7.56	7.42	7.28	7.13	6.99	6.86
25	8.27	8.09	7.92	7.75	7.58	7.44	7.29	7.15	7.01	6.88	6.85
26	8.12	7.94	7.78	7.62	7.45	7.31	7.16	7.03	6.89	6.86	6.63
27	7.98	7.79	7.64	7.49	7.32	7.18	7.03	6.91	6.78	6.65	6.52
28	7.84	7.65	7.51	7.36	7.19	7.06	6.92	6.79	6.66	6.53	6.40
29	7.69	7.52	7.38	7.23	7.08	6.95	6.32	6.68	6.55	6.42	6.29
30	7.56	7.39	7.25	7.12	6.96	6.83	6.70	6.58	6.45	6.32	6.19

Index

adjacency matrix 9, 23, 27
allometric principle 21, 37, 45, 118, 119, 120, 174
age structure 34, 99
ANN 5, 30, 36, 39
Arrhenius 55, 62, 75
artificial neural network 5, 30, 36

BCF 116, 119–20, 122, 126
biodegradability 76, 118, 122, 130
biogeochemical models 34, 45–82
bioenergetic models 29, 33–34, 39, 45–7
black box models 33, 36
BOD 66–7
buffer capacity 98

cadmium 121, 122, 126, 127, 129
calibration 15, 16–17
carrying capacity 89
causality 33, 39, 45, 47, 149
cellular automata 37–8, 141–3
chemo-state model 47, 48
chromium 135–40
classification of models 29, 30
complexity 19–21, 54
conceptual diagram 10–11, 45, 54, 60, 77, 113, 123, 170, 173
conjugation 144–50

3D-models 5, 30–1, 133, 134
decomposition rate 48, 75, 76, 80–2
Darwin's finches 167–168
DDT 112, 113, 130

diffusion 136
dynamic models 111, 112

Ecological Modelling (the journal) 2, 4–6, 31, 34, 37, 39
ecological niche 156
eco-exergy 159–62, 163, 168–9
ecotoxicological models 38–9, 109–29
EEP 116–18, 122
Eigenvalue 99
energy conservation 33, 95
environmental management 46
ERA 109, 115, 116–17
eutrophication models 4, 47, 49, 51, 54, 60, 63, 64, 172–4
exponential growth 12, 88

Fick's law 13
fishery model 1, 90
frequent measurements 21, 45, 60–2
fuzzy models 5, 30, 35–6

GIS 5, 30, 37, 133, 141, 143
Glumsø Lake 54, 60, 63–4
grazing 13, 54, 60
growth 57, 88–90

hierarchy 154
hysteresis 176, 177

IBM 5, 30, 31, 37, 38, 39, 133, 141, 144–50
IMM 22
individual based models 5, 30, 37–8, 133

infectious disease 102–6
information 73

K_{ow} 118, 119, 130
Kullback 161

Lake Fure 178
lake models 1, 58
Lake Mogan 172–6
lake restoration 180
lattice models 141–3; *see also* IBM
lead 112
logistic growth 13, 89, 90
Lotka-Volterra 3, 34, 87, 93–6

mass balance 9, 51, 112
mass conservation 29
maximum yield 92
mediated modeling 22–3
metapopulation 101–2
Michaelis-Menten 12–14, 24, 45, 55, 57, 60, 75, 80, 96
mixed flow reactor 48
model types 29–41
modelling procedure 8–19
Monod kinetics 12
mortality 88

natality 87
nutrient balance 179–81

oxygen consumption 74, 76, 80
oxygen depletion 26, 74, 75, 78

parameter determination 45
parameter estimation 116–19
parasites 168–72
parasitism 96
PCB 118
photosynthesis 10, 55
plant composition 50
population dynamics 34–5, 87–106, 113
prey–predator 94–6

re-aeration 75, 76
river models 73–8

SDM 5, 30, 35, 153–81
sensitivity analysis 14–15
sediment 51–2, 59
sedimentation 58
settling 51–2, 136, 138–9, 172
spatial models 37, 133–50
steady state models 34, 48
Stella 11–12, 47, 74, 76, 77, 79, 102, 104, 126–8, 154, 167, 169, 170, 172–8
stoichiometric approach 57
stochastic models 39
Streeter-Phelps model 3
submerged plants 173, 177, 178
subsurface wetland 47, 66
surface modeling 143–4
symbiosis 155

transparency 52–3, 63
two-steps description 57–8
tylosine 26–7, 123

uncertainty 21–2

validation 18, 19, 21, 63–4, 140
verification 8

wetland models 47, 84